W9-BMX-338

A FOSSIL-HUNTER'S NOTEBOOK

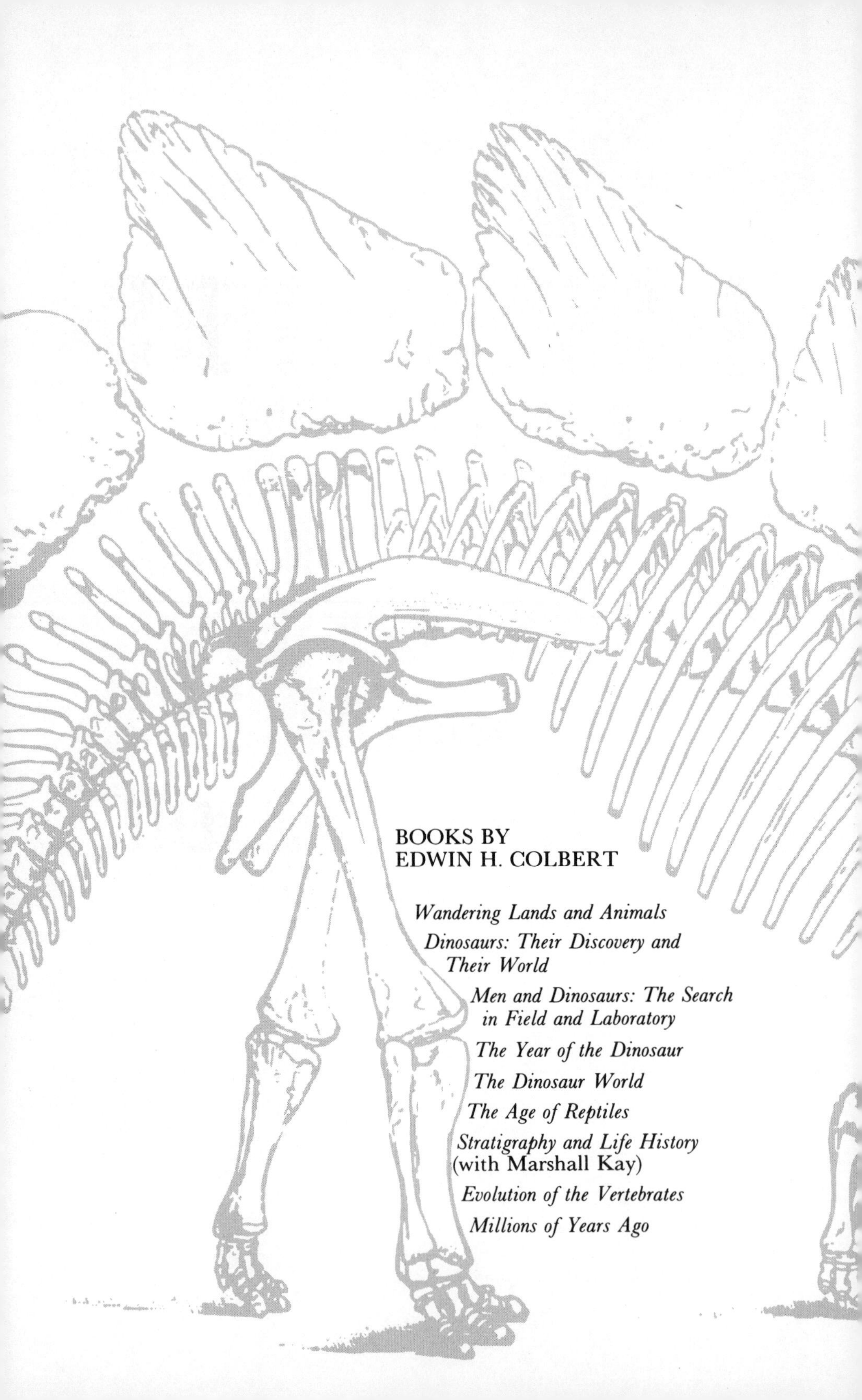

BOOKS BY
EDWIN H. COLBERT

Wandering Lands and Animals

Dinosaurs: Their Discovery and Their World

Men and Dinosaurs: The Search in Field and Laboratory

The Year of the Dinosaur

The Dinosaur World

The Age of Reptiles

Stratigraphy and Life History (with Marshall Kay)

Evolution of the Vertebrates

Millions of Years Ago

A FOSSIL-HUNTER'S NOTEBOOK

My Life with Dinosaurs and Other Friends

EDWIN H. COLBERT

E. P. Dutton · New York

All illustrations are from photographs in the possession of the author, except for the following.

Photos, pages 14, 16, and 29. The Missouriana Room, Northwest Missouri State University.

Photos, pages 68, 69, 71, 74, 115, 119, 123, 124, 127, and 129. Courtesy of the American Museum of Natural History.

Photos, pages 139 and 163. From *The Age of Reptiles,* by Edwin H. Colbert. (New York, W. W. Norton, 1966.)

Figure, page 156. Drawn specifically for this book.

Photos, pages 158 and 163. From *Wandering Lands and Animals,* by Edwin H. Colbert. (New York, E. P. Dutton, 1973.)

Photo, page 210. From *American Museum Novitates* No. 2535, by Edwin H. Colbert, 1974.

Photos, pages 218 and 221. U.S. Navy.

Photos, pages 233 and 234. The Museum of Northern Arizona, Flagstaff.

Painting, page 69. Collection, Columbia University in the City of New York.

For information contact:
Elsevier-Dutton Publishing Co., Inc.,
2 Park Avenue, New York, N.Y. 10016

Library of Congress Cataloging in Publication Data

Colbert, Edwin Harris
A fossil-hunter's notebook.
1. Colbert, Edwin Harris
2. Paleontologists—United States—Biography.
3. Vertebrates, Fossil. I. Title.
QE22.C54A33 560'.9 [B] 80-13261

ISBN: 0-525-10772-X

Published simultaneously in Canada by
Clarke, Irwin & Company
Limited, Toronto and Vancouver

Designed by Mary Gale Moyes

10 9 8 7 6 5 4 3 2 1

First Edition

To the American Museum of Natural History in New York,
my scientific home for forty years
and to the Museum of Northern Arizona in Flagstaff,
since 1970 my second scientific home,
this book is affectionately dedicated.

CONTENTS

ILLUSTRATIONS

A FOSSIL-HUNTER'S NOTEBOOK

1. EDWARDIAN YEARS, AND AFTER

I came into this world about one jump behind the airplane, on September 28, 1905, less than two years after the Wright brothers had made their first short flights on the sand dunes at Kitty Hawk. In the first years of my remembrance, airplanes, rare and flimsy contraptions that they were, occasionally adorned the pages of picture books and magazines but seldom were seen as three-dimensional objects.

One of the recollections of my early life, and it remains clear in my memory because it was so unusual a happening, was the visit of the first airplane to our town of Maryville, up in the northwestern corner of Missouri. I suppose the year was about 1910 or perhaps as late as 1912, and it was a big event for the entire county. Thousands of people were there. The plane was one of the early Wright biplanes of the type where the pilot, with his cap on hindside forward, the better to accommodate the large goggles over his eyes, perched on a little seat quite exposed on an open framework above a tricycle landing gear. The two canvas-covered wings were back of him, as was the engine, which transmitted its power by means of long, clanking chains to two pusher propellers.

The plane was to take off from and to land on a football field that in those days occupied a large, flat area behind the single building of the Northwest Missouri Normal School, a state teachers' training institute that recently had been established in Maryville. (My father taught mathematics at the Normal.) Of course the goalposts had been removed for the convenience and safety of the flyer, and the field was kept clear of people by many guardians of the lines—official and otherwise. But the crowd was so large that my parents took me into an upstairs room of the building, from which vantage point we could see over the crowd. Up the plane

That primitive plane was in the Paleozoic era of the internal combustion engine.

went—a marvelous sight—its white wings bright against the summer sky. It circled to the west over the undulating farmlands beyond the campus, and then it approached for a landing. We watched it glide in, and then we saw it rise again, as we heard the engine roar in a surge of power that carried it beyond the football grounds to an adjacent wheat field where it came down in a sudden and undignified landing. What had happened was that the crowd, unused to this new wonder of the air, had broken beyond the restraining barriers and rushed out onto the field to see the plane come in, and the pilot, to avoid a tragedy, had made a forced landing beyond the enthusiastic watchers. He was shaken up and more than a little bit disgusted, and the undercarriage of the plane was somewhat bent, which meant that he had to stay around town for some repairs.

I was taken for a close-up look at the slightly damaged plane—and what a disappointment it proved to be when examined at close range! The beautiful white wings were seen to be smudged with thousands of signatures, written on the canvas in pencil. The airplane had been a curiosity to those who had written their names on it, as it was to me, and it was to remain for several years an isolated curiosity in my young life. But perhaps the coming of that primitive plane to northwestern Missouri helps to place me in time—the time being the Paleozoic era of the internal-combustion engine.

I had been born in Clarinda, Iowa, about thirty miles to the north over the state line, but I was brought to Maryville before I was a year old, when my father accepted a position on the faculty of the newly created normal school. So Iowa was not truly a part of my early life. I grew up as a Missourian, but a northern Missourian having some, but not much, contact with the old southern culture that had been prevalent throughout much of the state.

We were a family of five—my parents, my two older brothers and myself. It was a rather extended family temporally speaking, because my older brother, Herschel, was fourteen years my elder, while my other brother, Philip, was nine years older than I was. Yet in spite of the age spread we formed a closely knit family with many bonds of affection holding us together.

When the family moved to Maryville my father was in his mid-forties, and when my memory of him comes into focus he was already into his fifties, a short, rather chubby, mild-mannered man with iron-gray hair and a clipped moustache. He had a rather round face, an ample nose, and light blue eyes that looked out over the tops of half-glasses shaped somewhat like orange segments in outline. In those days and for many years thereafter he was far-sighted; those peculiar spectacles, held on the bridge of his nose with a "pince-nez" spring (a fashion in glasses that seems now to be completely extinct), were a sort of trademark that was deeply imprinted on my early consciousness.

They say that mathematics and music go together, and my father was a musician of sorts. In his younger years he had been quite a chorister and had sung tenor in various musical groups. When I knew him he still had a nice tenor voice, and in our Presbyterian church he always sang the tenor lines, rather than going along on the main melody with the rest of the congregation. What I particularly remember was his flute, an old-fashioned black wooden flute with all sorts of nickel keys arranged along its length in addition to the regular holes. He would frequently get the flute out in the evening for a bit of practicing. The memory of one such evening is still with me. I was quite small, and I sat quietly and listened to his playing, and the pure notes of the flute brought tears to my eyes, young and musically inexperienced though I was.

A rather unexpected chapter of my father's early manhood that he recounted to me now and then was his experience as a baseball player. Of course he was an amateur, or at best a sort of semi-pro, but he had functioned as a shortstop way back in the days before players wore gloves. As a result he had a permanently crooked finger on one hand, the result of a hot drive caroming off the end of that digit. One could hardly imagine such activities on the part of so mild and placid a person.

It should be added parenthetically that my father may have been a

We were a family of five—a rather extended family temporally speaking, because my older brother, Herschel, was fourteen years my elder, while my other brother, Philip, was nine years older than I was.

mild and placid person, but he certainly was not in any way a timid type. He was not afraid to stand up and be counted and he did just that on more than one occasion. If there was an issue with which he was involved, controversial or otherwise, he would make his sentiments known in a most forthright manner.

As for my mother, she was anything but mild and placid. Hers was a mercurial temperament, and she often felt somewhat out of place living the life of a housewife in a small town. She would have loved to have been an actress (and indeed she did perform in amateur dramatics), or an author, or a world traveler. My father's idea of a summer vacation was to sit reading in a rocking chair at home, or something else equally restful; my mother's idea would be to go to the ends of the earth. But she never got the chance. Trips in those provincial days generally were restricted to journeys of a few hundred miles—to Colorado or to Chicago or as far east as Ohio.

Nevertheless, she made life interesting for herself, and for her family. During my early years something was wrong with my digestive system, so I managed by eating bland foods and spent quite a bit of time in bed. (I suddenly grew out of this condition about the time I reached high school.) Needless to say, it was my mother who watched over me. She was my good companion, participating in many of my activities and advising me in many ways. She would read to me and play games with me, and tell me fascinating stories about her hoydenish life as a young girl. (Certainly she did not always behave with the restrained deportment that was expected of a young Victorian lady.)

Herschel was enough older than I to seem like a surrogate father to me. He took a great deal of interest in me, and took me places when I was little, and did things for me and bought me toys. But when I was still quite young Herschel went off to the university, and after that I saw him only rarely, when he came home for visits. Phil was enough older than I was so that we were never playmates, but not enough older (as was Herschel) to take an avuncular role in my upbringing.

Since both of my brothers were so considerably removed from me by age they did things and had adventures that were quite beyond my limits. Thus they often had tales to tell—tales that I listened to with the wonderment and the envy of the uninitiated. How I yearned on more than one occasion to grow up, so that I could enjoy fascinating experiences such as they would talk about.

The stories that I heard from my mother and my brothers were always entertaining and much to be desired. But the stories I particularly liked were those told by my mother's father, my Grandfather Valentine Valencia Adamson. Grandfather Adamson lived with my step-grandmother in Holton, Kansas, where my mother grew up, and Holton was only a hundred miles or so from Maryville. Of course he and "Grandma" would visit us every now and then and we would go to visit in Holton. Grandfather Adamson had grown up on the western border, when that border was marked by the Missouri River between Missouri and Kansas. As a boy he watched the forty-niners get ready at Westport Landing (now Kansas City) for their long trek to the California gold fields. He knew Joe Robidoux, the proprietor of Robidoux's Landing (now St. Joseph). As a young man Grandfather went to medical school, and in 1859, soon after he had qualified as a doctor, he joined the Colorado gold rush and crossed the plains to Denver, when Denver consisted, in his words, of "a saloon, a blacksmith shop, a few huts and a collection of tents." He served with the Second Kansas Cavalry in the medical corps of the Union Army during the Civil War, and after the Civil War he practiced for many years as a country doctor in eastern Kansas. His stories of long ago would entertain me by the hour.

In 1859 Grandfather had teamed up with three or four rough and ready types for the westward trip to Colorado. After weeks of travel through the high prairie grasses, where they encountered great herds of bison (or buffalo, as these shaggy wild cattle were designated by American hunters and pioneers), and where they gingerly skirted encampments of Sioux Indians, they finally reached Denver.

Denver and the rugged mountain country immediately to the west had little to offer Grandfather. His search for gold was desultory and disconnected, and he soon became disillusioned with the whole business. So he decided to retrace his steps across the plains and set up a medical practice somewhere in eastern Kansas. It so happened that a troop of United States Cavalry was preparing to cross the plains from west to east, and the commander welcomed the opportunity of having a doctor along. Grandfather enlisted for the duration of the journey, and the troop set out in best Hollywood fashion, with equipments jangling and gunbarrels gleaming in the hot, bright sun.

One afternoon, several days east of Denver, they saw a lone prairie schooner some miles ahead of them, its white cover swaying back and forth as the wagon lurched along the rough track. The wagon also was headed toward the east. It seemed strange to the soldiers that an unaccompanied wagon would be crossing the plains—a tempting target for roving bands of horse Indians. But there it was.

Sure enough, at that instant a little band of mounted Indians dashed out from behind a small butte near the wagon, and immediately started to circle it—again in approved Hollywood style. The Indians were superb horsemen; each warrior clung to the off side of his horse so that he was almost fully protected by the mount's body. And from this precarious position each rider fired at the wagon from under the horse's neck. All the time there was a blazing rifle fire from within the wagon.

The fight was short. The cavalry, movie-style, put spurs to horses and rode pell-mell to the rescue. Long before they arrived the Indians had ridden off into a confusion of buttes and gullies, foiled in their brief attempt at plunder. When the cavalry came galloping up to the wagon they found inside a man and his wife, both wounded. He had a knuckle split by a bullet; she was transfixed by an arrow—the arrowhead protruding on one side of her body, the feathered end on the other.

Grandfather went right to work on her. He cut off the feathers, and pulled the arrow shaft on through her body. Luckily it had not pierced any vital organs, so Grandfather was able to treat and bandage her with some hope that she would recover. Then he turned his attention to the husband's split knuckle.

The pioneer wagon joined the cavalry troop and traveled east with the soldiers, now safe from attack. The man and wife, like Grandfather, had become disillusioned with the gold bug, and were trying to get back to familiar and serene surroundings. They made it without further incident. The wife quickly recovered and was up and about, helping with the cooking, before her husband's knuckle had healed.

That is the story. Nothing momentous, but a tale of life that now seems as remote as the Dark Ages. Yet I had it at first hand—plain and unvarnished.

My Grandfather Adamson was the only grandparent with whom I was acquainted; my maternal grandmother had died when my mother was a small girl, my paternal grandfather had died long before I was born, and my paternal grandmother died when I was very small. Moreover, Grandfather Adamson was the only family relative living reasonably close to us—ours was a scattered clan. There were occasional visits from uncles, aunts, and cousins, but for the most part we lived as a nuclear family, surrounded by friends but not by relatives.

We lived in a house that my father built on East Seventh Street in Maryville. It was the center of my private universe for a dozen and a half years, and a very nice place it was.

It was a reasonably ample house placed well back from the street on a lot more than four hundred feet deep. A simple box describes it, of white weatherboarding, crowned with a peaked roof, each slope of which was pierced by a dormer window, and with a flat-roofed porch across the front, or south side, continuing as an open porch around a part of the east side of the house. The porch was a pleasant place in the summertime and during the warmer months before and after the summer; ivy had been trained up along each porch pillar, along the railing and along the top of the porch between the pillars, to form a cool, green screen. There we spent many spare hours if the weather wasn't too hot, and in the late afternoons and evenings we would move around to the side porch where we could sit and talk and look out across the neighbors' yards, all of them, like ours, of large

Ours was a reasonably ample house, placed well back from the street.

size and filled with trees. It was rather like living in a park. And there we would hear the afternoon call of the peewee, a plaintive note, or in the late summer the insistent loud sawing sound made by hundreds of cicadas as the last light of day gave way to night. Then there would be the winking lights of innumerable fireflies.

Inside there was a long living room set transversely across the front of the house and behind it the usual dining room and kitchen. Upstairs were three bedrooms and a study, where my father had a commodious desk with a swivel chair in front of it. One corner of the study was my domain, where I had my "museum."

When I was about a dozen years of age my grandfather had shipped to me a glass-fronted case, perhaps three or four feet in height, supplied with shelves. I built a stand for it, and in that piece of furniture I set out my precious displays—complete with labels. They were mostly natural history objects; various local fossils of Carboniferous age, some nice trilobites, some minerals, a couple of birds' nests that had fallen to the ground (I wouldn't think of taking them out of the trees or of robbing them of their eggs), snake skins, and other assorted odds and ends. There were a few historical items, too: an old spur, some buttons, some coins and badges. It was truly an eclectic collection, but perhaps it was the beginning of a trend that was to decide the direction of my life. My father, an understanding and patient

LEFT:
Our porch was a pleasant place in the summertime.

BELOW:
It was rather like living in a park.

man, put up with this idiosyncrasy of mine in a remarkably tolerant manner.

Back of the house was the larger part of our deep lot, containing various fruit trees and garden plots, a grape arbor, behind that a barn, and behind the barn a potato and corn patch. Beyond our back boundary was a big field, and on the far edge of the field was the line of the Wabash railroad. We could look out and watch the trains, which were just pleasantly distant. It is said that when the family first moved into the house, they occasionally would haul me out of bed at eleven, to watch the express train go through from Omaha on its way to St. Louis. It was a wonderful sight then, and it would be a wonderful sight even today if it could be re-created. Down the grade would come the train, with the headlight piercing the black night, and with the red glow from the open door of the boiler (the fireman was always stoking up at that stage of the run) casting its light up on the smoke streaming back overhead from the stack. Then there was a dark part of the train—the baggage and mail cars. After that were rows of bright windows which were the chair cars, and at the back of the train irregularly spaced squares of light in the Pullman cars, the lights of travelers who had not as yet gone to sleep. Away went the train out of sight, and we could hear the long moaning whistle for the grade crossing east of our house. It was a magic sound. The old steam whistle had a melodious note, and in the middle of the night one would wake up just a little bit, to hear through a delicious drowsiness the call of the train across the dark distance. Our modern world is poorer for the disappearance of this haunting sound.

That single-track railroad line, sometimes elevated on a low embankment, sometimes hidden in cuts, was the back boundary of our neighborhood. Beyond it was another large field traversed by a brook, where in the spring we—the neighborhood children—would gather violets and other wildflowers, which we would take home in triumph and with some excitement as presents for our mothers. And then, on the other side of the field, were some scattered houses forming the outer northern limits of the town. With the deep lawns in front and the railroad with its attendant fields in back, we were to some considerable degree set off by ourselves, a fact that gave our little group of houses and their inhabitants a certain amount of physical and psychological cohesion.

Next to us on the east was the Roseberry house, where lived Esther, a year younger than I was, and a constant companion of my very young childhood. As little children we played together, occupied hours on end with the childish interests which very small people enjoy. To us the railroad was an enduring backdrop for our little world. We were impressed, so much so that we made our own trains out of shoe boxes, hooked together with string, which we pulled across the grassy lawns for hours at a time. And beyond the Roseberry house was the Frank house, where Esther's cousin, Eva

Margaret, two years my junior, lived.

On the other side of us, several houses away, was a modest home occupied by the MacPherron family, a moderately numerous tribe most of whom were older than I was. But the two youngest boys, Edwin and Ora, particularly Ora, came more or less within my age bracket; it was with Ora that I spent many boyhood hours. He was always known to us as "Little Mac" to distinguish him from Edwin, who was "Big Mac."

Little Mac was full of ideas, and that made him a lot of fun. With my far from expert aid he built boats and little wagons and other objects dear to the hearts of boys. He once built a cannon, consisting of a piece of gas-pipe nailed to a heavy block of wood. One end of the pipe was solidly plugged up, and near this end he filed a notch so that it communicated with the interior of the pipe. He would partially fill the pipe with gunpowder, ram a piece of wadding in to hold the powder, and then sprinkle a little powder on top of the pipe so that it dribbled down through the notch. Then we would get set to jump back, Little Mac would touch a match to the loose powder, and before we could retreat more than half a step the cannon would detonate with a loud and very satisfying roar, sailing into the air and turning a couple of somersaults as it fired. That furnished a sort of diversion for idle moments.

But Little Mac's masterpiece was his bean thresher. Before moving into the house near us the MacPherrons had lived on a farm, and Little Mac was acquainted with farm machinery. He understood the principle of a threshing machine, so he decided that we should build a small-sized version, to take the hulls off beans or peas. It consisted of a frame, within which was arranged the head of a croquet mallet studded with nails, the heads of which had been filed off, and mounted on a horizontal axle, driven through its long dimension. The idea was that the croquet mallet, when activated by a pulley attached to one end of its axle, would go around and around, the array of rotating nails making a formidable shredding device. Beneath the croquet mallet was a movable screen, operated by an eccentric, so that it shook back and forth. The screen was tilted, and beneath it was a box, while at its end was another box. A large grindstone was set in place and aligned with the pulley on the croquet mallet. A heavy cord was passed around the grindstone and around the pulley, and made tight. Then the grindstone was turned and things happened; the mallet rotated at high speed, the screen shook back and forth, and the whole contraption made a wonderful lot of noise. We fed the beans or peas into a hopper (which I should have mentioned) placed above the croquet mallet, the vegetables fell down and got shredded, the unencased beans dropped to the shaking screen and fell through its meshes to the box below, while the shredded hulls were shaken down the tilted screen and dropped into the other box. That was the theory. In practice bean hulls flew right and left, and many beans ended up in the trash box instead of where they belonged. Then we would have to pick over the dump, as it were, to rescue the errant legumes. Nevertheless, we had a lot of fun shelling beans or peas for our mothers, and it probably didn't take more than twice as long as if we had done it by hand.

One of the golden days of my childhood was when Mr. MacPherron drove to Stanberry, a small town about thirty miles from Maryville, and took Little Mac and me along. It was a perfect autumn day, a dreamlike day when the sun was warm but just a bit hazy, the trees were of varied hues, and the corn stood ripe in the fields. We drove over the narrow, dirt roads in the MacPherron Overland, a species of car long since extinct, and we took the day to do the trip. At noon we had a picnic lunch by the side of the road, and never did food taste more delicious. Mr. MacPherron did what he had to do in Stanberry while Little Mac and I enjoyed the warm midday and afternoon sunshine. Then home in the late afternoon as the shadows grew long and the air grew chill.

That's all there was to it. Nothing exciting, nothing unusual. But it was such a beautiful day, and so pleasant a trip that it has remained imprinted with remarkable clarity upon my memory through more than six decades. When I think back to some of the pleasures of being a small boy, I think of that fall day of sunshine and autumn haze, of brightly colored trees, of corn waiting to be harvested and orchards full of bright red apples, and of the delicious pastoral smells of a countryside completely innocent of industrial noise and pollution.

Beyond the Roseberry and Frank houses on the one side and the MacPherron house on the other there were the less intimate reaches of the town. Yet these areas outside our neighborhood were not particularly strange to me because ours was a small town; all of its sections were familiar even when I was little, so that I could venture forth into near and distant neighborhoods by myself. Three blocks, and not very long blocks at that, would take me from our front steps to the edge of the business district. And four or five blocks would take me across the business part of town and into the residential section on the other side. Getting around town therefore was not much of a problem, and it was something that could be achieved by a boy even at a fairly tender age.

Some people owned cars, but such cars as there were, were primitive and varied. The Roseberrys had a chain-drive Buick—a noisy contraption, and Mr. Todd, around the corner, had some kind of car that cranked up on the side instead of in front. In those days before self-starters all cars had to be cranked by hand, and there was an art to cranking a car engine. The bigger the car the heavier the engine, and the harder it was to "turn over." So one had to know just how to set the gas and the spark, and how to hold the crank and the most effective way to give it a turn. Half a turn or a single turn was usually enough, except when the engine was cold and stiff. But it had to be a good, strong, masterful turn, with the thumb held parallel to the other fingers, *not* wrapped around the crank handle. Many a strong man in those ancient days of the automobile had his arm broken by the crank "kicking back" because of compression in the cylinders, and such broken arms were usually the result of a too firm grip on the crank handle. With the thumb held alongside the other fingers, the crank would kick out of the cranker's hand, and thus no harm would come to him.

The Gillams, who lived a couple of houses away from us, had a Cadillac and cranking that car engine was a job for a strong and determined arm. Those were the only cars in our immediate neighborhood. The Corwins, family friends who lived on the other side of town, had an electric car, as did Mrs. Colby, one of my mother's friends. There were other electric cars in town, and they were very convenient for getting about in, much more convenient than the gasoline motors of those days. The electrics looked like tall china cabinets on wheels; they had short wheel bases, solid rubber tires, and were remarkably simple to drive. The driver had before him or her a horizontal bar for steering—forward to go left, back to go right (I think that was it), and on the end of the bar was a little control to turn the power on or off. There was a foot brake. And that was all. Those old cars whirred along at a constant speed, up hill or down, they were almost silent, and they were pretty safe. They would go only about twenty miles or so on a single charge, but that was ample for a day's journeying around and about our town. At night they would be plugged into a big socket in the garage, to renew the charge on the battery; then they were ready to go—instantly.

Even though automobiles were becoming increasingly numerous and sophisticated with each passing year, horses were still the old reliables during those Edwardian and early post-Edwardian days. Beyond the town limits the roads were all dirt—hard and dusty in fair weather, veritable quagmires after heavy summer rains and all through the long winters. Out on the farms horses pulled the plows and harrows and rakes, and on Saturdays they brought the farm families into town for the weekly shopping.

On three sides of the courthouse square were long files of heavy posts, connected each to the next by a stout chain for the hitching of horses. On a Saturday this was a crowded spot, and the red-brick courthouse, with its tall tower, had the appearance of an equine shrine, with horses of all colors hitched to buggies and wagons, facing it from three of the four cardinal points of the compass. The fourth side of the square, on Main Street, was kept unimpeded.

It was on that side of the square that one day I saw a man at the wheel of a Model T Ford (evidently he was an inexperienced and terrified driver) mount the curb with his right wheels to zoom along with two wheels on the sidewalk and two wheels in the street. His progress inevitably came to an abrupt halt when he collided with the lamppost on the corner, causing it to sway back and forth like a tree in a high wind and shake off the five globes that crowned it. Old Cy something or other (I forget his last name) was just then coming along in his spick-and-span open carriage, pulled by a very sleek and well-groomed horse. The lamp globes came down on the horse, and immediately there was an urban runaway, with Cy hauling on the reins and doing his best to stop the frightened beast. The horse did not run far, and nobody was hurt. It was a graphic and comical example of the impact of the automobile age on an earlier form of transportation.

Deliveries to our house were by horse-drawn vehicles: a rather large, enclosed wagon that daily brought the ice; a light wagon, pulled by two small horses, that brought the milk; a spring wagon that delivered the groceries; and in the fall, a heavy wagon, pulled by two massive horses, that delivered our winter's supply of coal. Of course for anything as crucial as fire protection it was horses, a pair of strong, shining horses that charged through the streets pulling the town's single fire engine, a hose and ladder wagon. It was a thrilling sight, the horses straining at their harness and running in unison, striking sparks from their iron-shod feet, the fire wagon rattling over the brick pavement, and the driver sawing on the taut reins.

Horses of all colors hitched to buggies and wagons.

In between these tense few moments of racing the horses occupied separate stalls at the back of the firehouse, where they stored up energy on an ample diet of oats and hay, in preparation for the next emergency.

Across the street from the firehouse was a blacksmith shop, cool and dark inside except for the light from the red fire burning in the forge over in one corner, and always emanating a strong, horsy smell. I remember a day when I went to the open door of the shop, firmly holding to my father's hand, and watched the blacksmith pull a horse's hoof up between leather-aproned legs, to fit and attach the shoe. And I wonderingly asked my father why it didn't hurt the horse to have nails driven into its foot. He explained that the blacksmith drove the nails only through the edge of the hoof, which would be like drilling a small hole through the free edge of a fingernail. Then I understood.

Outside the shop there might be several horses standing hitched to wagons, some of them waiting their turns to be shod. These horses stood still because a rope or a long leather trace, attached to the bit in the mouth of each horse, was fastened at the other end to an iron weight about the size of a small teakettle. Why, I asked my father, did not the horse just walk away, pulling the weight with him. Again he explained things to me; he told me how the bit was across the tongue of the horse, and how, if he tried to pull at the weight the bit would press down on his tongue with sufficient force to make him stop. The tongue was tender, he said, just like my tongue. Again I understood, and one more item was added to my early education.

An event that introduced some excitement to my early life was when Buffalo Bill's Wild West Show came to town. That vignette remains rather clear. My mother took me to see it, and my joy was unalloyed, except for the moment when a balloon, which she had bought for me, exploded. I was unhappy, and of course she was sympathetic, but the balloon was soon forgotten when I saw people ride out on horses and at a fast run shoot down glass balls that were thrown into the air by accompanying horsemen. There was other trick shooting (I wonder if I saw Annie Oakley) and trick riding. The high point of the show was when William Cody rode out to the center of the field on a white horse, a man in white buckskins, with long white hair and neat whiskers. He took off his wide-brimmed hat and saluted the cheering crowd with practiced aplomb. It was a sight to be remembered.

It was perhaps a couple of years or so after Buffalo Bill came to town that I advanced from the unfettered status of a very small boy to the somewhat regimented position of a not quite so small scholar. My grade-school years were spent largely at the "Training School," which was housed on the ground floor of the Normal building, out on the west side of town. And it was here that I truly entered the world of boys of my own age, separated during school hours from my former neighborhood playmates.

The "Training School" was housed on the ground floor of the Normal, out on the west side of town.

Among my several classmates and playmates of those days were the two Landon boys, Truman and Kurt. Truman, the elder of the two brothers, was an especially close friend of mine. Their father was Perry Oliver Landon, who ran a conservatory of music in town in the halcyon days before the First World War. P. O. Landon had gone to Germany to study music and had married a German woman during his sojourn abroad. Truman and Kurt grew up speaking both English and German, and that impressed me. Also the Landons raised pedigreed dachshunds, which made visits to their house particularly interesting.

Both boys, when they grew up, went to West Point and became professional soldiers. During the Second World War, Truman was a general in the Air Force, and after the war he became a four-star general, promoted by President Kennedy. For some years in the fifties he was in command of the American air forces in western Europe.

One day when Truman and I were in the sixth grade, I think it was, we looked out our classroom window and saw a strange object in the sky. It was a round balloon floating along a few hundred feet above the ground. One did not see free-floating balloons very often in those days (indeed, they are in most places something of a novelty today) so the teacher excused us all from class to go out and get a good look at the strange object. Truman and I ran along, waving at the two occupants of the basket, and they waved back. It was a thrill for me, and I am sure a thrill for the future Air Force general. We learned later that the balloon came from the Army Signal Corps station in Omaha. It was on some sort of a practice run (do balloons run?).

I had another experience with a balloon in those early years. It was at the County Fair, when I was perhaps eleven or twelve years old. During the fair a balloonist entertained the crowd. A fire was built on the ground inside the racetrack, to create hot air to inflate the balloon. How they got things started without setting fire to the balloon I do not know. At any rate, it was done, and the balloon began to take shape. Finally it was inflated, and up it went with the balloonist underneath, not in a basket but on a trapeze. As the balloon rose into the sky, he performed tricks on the trapeze, much to the amazement of the assembled crowd. Then, at a considerable altitude he pulled a rip cord, the balloon deflated, and he leaped out and descended to the ground via a parachute.

My friends and I, seeing the direction of his descent, started to run toward his projected landing place. The most practical way for us to get there was to run along the right-of-way of the Burlington Railroad, a branch line that ran from Creston, Iowa, to St. Joseph, Missouri, crossing the Wabash a half mile or so to the north of the fairgrounds. Down the track we went pell-mell, and around a bend we came upon the balloonist, sitting on a trestle, his parachute draped over the rails and ties in front of him. He seemed a bit dazed; I think he must have had a rather hard landing. He sat there for a while, and we stood around respectfully and asked him questions. Finally he got to his feet, gathered together his parachute, and we all walked back together along the track.

The County Fair was one big summer event. The other was the annual Chautauqua. The original Chautauqua was, of course, the institution established in the expansive years of the late nineteenth century on the shores of Lake Chautauqua, New York. Naturally a majority of the far-flung population of the country could not make it to Lake Chautauqua, so lesser "Chautauquas," as they were called, sprang up and evolved across the land. Most of these were not of the fixed variety, but rather were traveling lyceums (if that is the word to use) that went from town to town during the summer.

First a crew would come to town on the train, packing with them an enormous tent, to be erected in a suitable spot. These tents could accommodate several hundred, perhaps a thousand or more people. I suppose the crew brought the seats with them, too, as well as a stage and other accoutrements. The stage would occupy one side of the long dimension of the tent, and the seats would be arranged in extended, semicircular rows facing the stage. The stage had a curtain in front and canvas dressing rooms at the back. Long strings of electric lights were strung up within the tent, so that there could be evening performances as well as daytime programs. When everything was in order, the performers and entertainers and lecturers came in succession, day after day, to present their programs. The scheduling of these small-town Chautauquas staggers the imagination, looking at the phenomenon in hindsight. For it must be realized that Chautauqua tents dotted the countryside—every fifty miles or so in the Midwest, and the performers made their way daily from one town to the next. In those days there was no such thing as making the circuit by car; travel necessarily was by train, and trains did not always run at convenient times. The people who traveled the Chautauqua circuits must have had iron constitutions and untold reserves of energy.

When I was small the Chautauqua was held in a large park on the Normal School campus. The big tent was pitched on a slight natural declivity so that the files of seats rose, one behind the other, much as in a theater. All of which was very convenient for the audience. A dirt road skirted the back of the tent, allowing access to the grounds by horse and carriage and by primitive automobiles. On the other side of the road, set in a pleasant grove of oak trees was a tent city, occupied by many of the more enthusiastic Chautauqua-goers.

People in town would rent large tents, or perhaps buy them, and set them up on the Chautauqua grounds. They were spaced in an orderly manner so that grassy lanes ran in front of and behind each row of tents. Here many families would spend their days. They would bring picnic lunches and dinners, so that they could be at hand for the afternoon and evening programs, as well as the morning events for children. The tents made pleasant places in which to rest and visit during the intermissions. Moreover, some people brought cots, and even spent the nights, camping out in cultured style.

So the long, hot midwestern summers were punctuated by a week of learning and entertainment, in an age that now seems remotely distant and in many respects rurally innocent. Most of the programs at our Chautauqua were not absolutely first class, but then they were not exactly mediocre, either. They brought the outside world to us, and they presented it in the round, so to speak, which gave us more sense of participation than looking at the tube does today.

The Chautauquas persisted until the early twenties in our part of the country. Then they faded away as a result of the impact of motion pictures and radio.

In our town we were occasionally visited by the great and the near-great at times other than Chautauqua week. One of my early memories is of President William Howard Taft coming to Maryville; I think when he was campaigning (unsuccessfully) for his second term. I did not see him in town, but I went with my family down to the Wabash railway behind our house, to see the Presidential Train come past on its way to Saint Louis. We waited expectantly and in due time down the track came the pilot engine, a lone locomotive that preceded the Presidential Train by a couple of minutes or so, to be sure that the track was clear. (The gruesome tactic was that if someone had planted a bomb on the track the pilot engine would be blown sky-high, but the President would escape.) Then came the train, and on the back platform was the President, a jovial, moustachioed fat man of huge dimensions. We waved at him as the train went past, and he waved back, the only time I have ever been greeted by a President.

Some four years later Vice President Thomas R. Marshall came to town, during the campaign of 1916. He and his party drove out to the Normal in a dreary autumn rain, where he gave a brief talk. I was in the audience. I don't know that I absorbed much of what he said, but I did look upon him as an old acquaintance; some years earlier he had attended the wedding of my cousin, Katherine Croan, to Walter Greenough, in Anderson, Indiana, and I, dressed in white and functioning as a sort of miniature usher at the wedding, had shaken his hand and talked with him.

I had traveled to Anderson with my mother—of course by train, that being one of several train trips during my early years. Children today, riding by car across the continent and contemplating the passing scene with bored indifference, can hardly realize the excitement of a train trip in those great days of railroading, especially the joy of riding on a Pullman. There we were, in semi-secluded grandeur, rollling across the land on safe, steel rails. One of my recollections is of my mathematical father listening to the click of the wheels on the rails while looking at his watch. He knew the length of the rails, he counted the number of clicks per minute, and from that he calculated the speed at which we were traveling. I was filled with awe.

Then there was the long walk to the diner, and a pleasant meal, with snowy white linen and gleaming silver and attentive waiters. Ah! Those were the days to travel—*if* (I may say) one were middle class and white.

Then back to the Pullman, and the berth was made up. I especially liked the little hammock that stretched along the windows, in which one could stow odds and ends before turning in. Finally, there were the delicious nights, the bed rocking with the motion of the train, and the lone whistle, sounding its magic note far, far ahead.

In such a world—a small-town world surrounded by a verdant landscape of cultivated fields, a world of friendly neighbors, of summer Chautauquas, of family trips on trains—I was quite unexposed during my very young years to some of the harsher aspects of life. But even in our countrified environment there were problems which in the course of time intruded upon my consciousness.

The first time that I was brought face to face with a sometimes hard and bitter world, far removed in spirit if not in distance from the protected environment of our family home, must have been when I was no more than four or five years old. That was the occasion when my mother went with our neighbor, Mrs. Gillam, to visit the county Poor Farm, and I was taken along. We went in the Gillams' car and were driven by a local worthy, hired for that day to perform the duty of a chauffeur. As we drove along the country road to the Poor Farm, I wondered what kind of place it might be; I could not picture it in my imagination.

In due time we arrived at an institution consisting of a number of brick buildings set back from the road in the midst of some very ordinary cornfields. We got out of the car, leaving it in the care of the driver, who parked it beneath the welcome shade of some large elm trees, and went up the steps of the main building into a sort of reception room. There we were met by the Superintendent, I suppose it was, or at least some person of authority, and under the guidance of this official we were taken around to several of the buildings of the Poor Farm.

Of course such an institution in those days was largely for the accommodation of destitute elderly people, and those were the people we saw. They were old, sometimes rather helpless, and universally miserable. I was truly depressed by what I witnessed, but it was not the helpless old age of the inmates that affected me, it was their evident unhappiness. Their condition struck me, small and innocent as I was, with traumatic impact. I could not erase from my mind the sight of one old lady, abandoned by her two sons (so we were told), sitting forlornly in a rocking chair and staring fixedly at nothing. I don't think my mother was quite aware of the effect the visit had on me.

We left the Poor Farm after being there a couple of hours, and all the way back to town I was almost overwhelmed by what I had seen. When we got home I asked my mother all about it and she tried to explain the situation to me. But the experience loomed large in my consciousness for a long time; the memory of it is still clear in my mind. I had glimpsed an unkind world—a world that hitherto had been completely out of my ken. For a very small child came a realization that there exist large and frightening problems, problems of such dimensions that little childhood worries seem for the moment insignificant.

The big problem in Maryville, as of course in almost every town and city in the nation, was that of race, specifically the problem of blacks and whites. We did not have a large black population, but few and humble as they were, they were segregated. Most of them lived in the east end of town, and all of them were carefully separated from the whites in the school system (they had their own school, a poor little frame building in which the elements of learning were set forth in a much more meager way than was the case in the white schools), in religious life (they had their own two churches), and in public gatherings (they sat in the balcony of the Empire

Theater, where old-time silent films were displayed). Needless to say the black children did not participate in the organized activities, such as the Boy Scouts. This segregation was accepted by all—unthinkingly by the whites, silently and with an undoubtedly inward rage by the blacks. In short, we had a race problem but we didn't seem to know it.

As for me, I was one with the rest of the white population in that I hardly gave the problem of black and white a second thought. I accepted segregation as it existed in our community as one of the inevitable facts of life and did not become truly aware of its heinous nature until I had grown up and left home. I was in more or less the same situation as was Mark Twain during his boyhood, for he states in his autobiography that: "In my schoolboy days I had no aversion to slavery. I was not aware that there was anything wrong about it." Children are necessarily imprisoned by their environments, and they live according to the rules by which they are governed.

Aside from the ever-present problem of black and white, our town was not especially plagued by problems. It was a generally tranquil little county seat, far removed from most forms of social unrest. There was virtually no manufacturing, so there were no real labor problems. The population was generally homogeneous; among the whites there was no foreign minority. The people were largely descended from original settlers of northern European stock. The only language I ever heard was English, or at least the midwestern variety of American English. The townsfolk and the country people as well were what simplistically might be called "pure American."

Yet they were not exactly WASPs, either. White they were, and Anglo-Saxon in a broad interpretation of that term, but certainly they were not overwhelmingly Protestant, for Maryville had a very large Roman Catholic population. There were nine churches in town when I was a boy, five white Protestant churches, two black churches, and two very large Catholic churches, St. Patrick's, the "Irish" church, and St. Mary's, the "German" church. In addition there was a Catholic hospital and appended to it was a chapel. Fourteen miles away were two large Catholic institutions, a Benedectine monastery and not far from it a convent.

This strong admixture of Protestants and Catholics was completely friendly, so far as I can remember. One of my closest friends, Paul Diss, was a Catholic. And among the Boy Scouts the Protestants and Catholics were about equally balanced: I know, because there were two troops in town, one made up of Protestant boys, the other of Catholics. Perhaps that might be called segregation of sorts. However, the two troops worked together and met together and went to camp together and the boys mixed with each other, seldom thinking about any problems of religious differences.

Looking back, those distant days seem like a tranquil time in a generally peaceful corner of the country. It was a time when people where we lived were not particularly interested in world events. Most of their worries had to do with local issues, such as they were.

And there did not seem to be many of the problems that worry us today. Pollution was confined largely to jokes about the black smoke that came out of the chimneys of Pittsburgh. The population explosion was a subject for a future generation. There were few worries about the environment; it was there as it always had been. Nobody worried about highways cutting across the countryside, because in those days paved highways were things much to be desired. We were just then pulling ourselves out of the mud and rejoicing at the disappearance of the deep ruts and the dust that universally characterized roads going up and down hills, along section lines in a mile-square grid that covered the land. The sound of the airplane was not heard in the sky. It was a simple time, an unhurried time. In such a time our problems were for the most part of four-cylinder, low-horsepower magnitude. They did not roar through life as they do today.

2.
GROWING UP

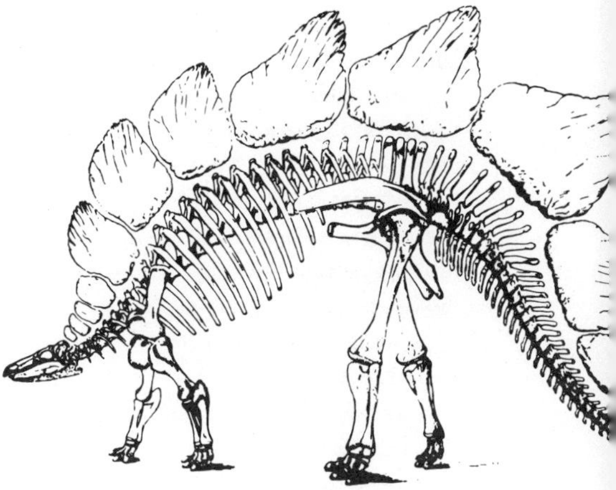

It was on a bleak February afternoon in 1917 that my mother came into the house (she had been downtown) and announced that our country had broken diplomatic relations with Imperial Germany. She was in a very discouraged mood that I could not quite understand, primarily because I did not understand what it meant for one country to break diplomatic relationships with another. Since the news didn't seem particularly disturbing to me, I asked her why she was so sad. Then she sat down and explained the matter, and I began to realize, after a fashion, why she was feeling blue. The break in diplomatic relations signaled almost certainly our entrance into the Great War that was raging across Belgium and France and northern Italy, and in our family there were two sons who almost certainly would be involved in that conflict. That seemed to be the shape of things for us during the months to come.

In May, soon after the United States declaration of war against Germany, Herschel entered the first Officers' Training School at Fort Sheridan, Illinois, not far from Chicago. He had been a four-year ROTC man at the University of Missouri, so he went to camp with a certain background of military training, and he came out of camp, ninety days later, commissioned as a captain. He came home for a visit and was with us for about a week. Then he went off to war—not to France but to Camp Bowie, Texas, where he was detailed to train officer candidates, and that led to my most memorable war experience.

In the spring of 1918 his departure for France seemed imminent, so he wrote to my mother and asked her to come to Texas to see him before he went abroad to a very uncertain future. But she couldn't go away and leave me, so she thought. Well, said Herschel, bring me along and he would pay my way.

It was a wonderful prospect for me, a trip all the way to Texas, and best of all, a three-week holiday from school. We made our plans, and early in March we boarded the train for St. Joseph, where we changed to another train for Kansas City, where we were to catch the Missouri-Kansas-Texas "Katy Flyer" for Fort Worth. In due time we arrived, and Herschel met us at the station.

Camp Bowie was a sprawling agglomeration of hastily erected wooden barracks and tents on the outskirts of Fort Worth, where some 60,000 men were training for overseas duty. It so happened that there was a little isolated suburb consisting, as I remember it, of only a few houses along a single street that had been surrounded and engulfed by the great military camp. And in one of these houses Herschel had arranged for a room for Mother and myself.

For me each day was an adventure—watching men marching, watching men at bayonet practice, watching men leaping out of trenches to charge across open fields. Then one day I got to participate in a maneuver. Herschel and some other officers assembled several hundred trainees in front of the barracks, and after some lecturing and instructions we went off along dry, dusty roads into the countryside. Certain men in the Training School had been selected as commanders; Herschel and the other officers went along as observers, and of course I tagged along with Herschel.

The roads were deep in white dust, derived from the Cretaceous rocks of north-central Texas, so that before long we were all thoroughly coated with the dust, and the khaki uniforms had become very light colored. Perhaps that was good; it made the men blend in with the dusty landscape. As we made our way along the road where it bordered a little stream the world suddenly exploded in a cacophony of rifle fire. We were ambushed!

The men all hit the dirt and started to fire back. This went on for some time, and then a charging line of men came at us through the trees and the dusty grass. Our side got up and charged, too, but just as the two lines of men were about to collide in mortal combat they all stopped.

Then the officers came forward. The men sat down, and there was a lecture and a discussion from the officers as to what had been done right and what mistakes had been made. After that the two forces joined, and we all marched back to the barracks, ever getting dustier as we went along. It was the only battle in which I have ever participated, and I guess it might be classed as more of a skirmish than anything else. For me it was exciting enough.

All too soon for me our visit was ended, and we boarded the train for the more mundane world of northwestern Missouri. (As it transpired, Herschel never did get to France. The army, in all its wisdom, decided to keep him in Texas to train more officers. He was a good teacher.)

I might add a few words about Phil's wartime experience, which was difficult and at times traumatic. He went to the Missouri School of Mines for a year, and then he tried to enlist. He was not accepted, but he was caught by the draft. He wanted to serve in the engineering corps but was denied that possibility. So he went to Texas as a buck private. For a time he and Herschel were in the same camp, Camp MacArthur, to which Herschel had been transferred, but they had little opportunity to see each other.

In the fall of 1918 the great worldwide influenza epidemic struck this country, and the army camps were not exempt. Phil was yanked away from his unit and assigned to a camp hospital as an orderly. For weeks he worked in the wards, seeing men die right and left. It was a bad experience, and he was more than happy when the Armistice came, with the result that within a short time he was released from the service.

But to go back to my own life during the war, it was spent largely within the confines of the seventh and eighth grades. Learning had to be continued, although it was interspersed with patriotic songs, rallies, and the like. Outside of school hours there were other things to do, especially those activities in which the Boy Scouts participated. I had become a Scout in 1917, after having reached my twelfth birthday, and for several years I was an enthusiastic member of that organization. Scouting in our town was not the structured affair it is now. It was loosely organized and loosely administered, and everybody had a good deal of fun. We didn't feel any pressures to advance through the hierarchy of scouting, although I did manage to become a First-Class Scout during my membership in the movement.

With the war on, various things were found for the Scouts to do. We knocked on doors and urged people to buy War Bonds and Stamps. In fact, I eventually received a bronze medal (long since lost) for helping to sell War Bonds. We collected scrap metal. I remember one time when we all went out in the woods and measured black walnut trees. The idea was to build up an inventory of black walnut trees which might be used for the manufacture of airplanes, especially propellers. And for some weeks I was part of a two-boy detail that appeared every morning and evening at the intersection of Main and Fourth streets to raise and lower the flag. In a burst of patriotic fervor a flagpole had been erected in the middle of the street at this place, constituting a visible symbol of the town's sentiments and at the same time something of a traffic hazard. However, cars were sufficiently slow and far apart so that it was not much of an obstacle.

On November 8, 1918, I heard the church bells ringing all over town. I think it must have been a Saturday, because I remember not being in school. Then people were shouting and there were the reports of guns being fired. Soon the word went around—the war was over! I ran downtown where a scene of a pandemonium met my eye. People were in the streets shouting and calling to each other. Many men had their hunting guns, shotguns and rifles, and were firing them in the air. Cars were sounding their horns. Everybody was supremely happy. And as I ran through the crowds with my fellows, picking up discarded shells from the guns, I could now and then feel the rain of buckshot descending from the skies. (I suppose the men with rifles were firing blanks—I hope so.)

Alas, it was all premature; it was the famous "false armistice," celebrated throughout the world. Negotiations were under way for the surrender of the German army, and in some way the news of the proceedings had been wrongly interpreted as an actual conclusion of an armistice. The Armistice was concluded three days later, at eleven in the morning, on November 11. Of course eleven in the morning in the Forest of Compiègne, where the Armistice was signed in a railroad coach, was a bit before sunrise in Maryville. Thus the town had a chance to celebrate the true Armistice, which was done with a parade and speeches. But it was all a bit anticlimactic and lacked the spontaneous enthusiasm of the previous celebration.

The war really was over! Its beginnings are very dim in my memory, but its final two years stand out in my mind with great clarity. In those years between the beginning and the end I had grown in stature and perception, while during the closing months I had participated in my own very small way in a rush of events that had overwhelmed people of many persuasions across the face of the earth.

In a sense our participation in the war formed a turning point of sorts in my young life. During those years leading up to the war, and then in the following two or three years before our country was fully involved in the conflict, I was a little boy "living without thought or care, loving only to play," as was said by Bartholomew of England, the thirteenth-century encyclopedist. With our entry into the war, the service of my two brothers in the army, and my own small efforts at home, I quickly developed a feeling of some responsibility toward the world in which I lived. I could see, even if dimly, the more serious aspects of life. In addition, the end of the war coincided roughly with my transition from grammar school to high school, with my entry into a new and perhaps a more adult pattern of living. Thus my years in high school were years of growth—growth in my perception and understanding of the world in which I lived, and physical growth as well. It was during those years that I attained something of the stature befitting a teenage young man. Of course such is the usual pattern for boys, but in my case it was accentuated because I started high school as an undersized lad (the smallest boy in school) and then in the course of the next four years reached a height more or less in accord with that of the other boys in my class. I never became large, but I did manage to grow to a little less than average size, and that was a great boost to my ego.

Maryville High School was a comparatively small institution, with about three hundred students, so everybody knew everybody else. The school was pretty much in the middle of town, and we all walked to school. There were a few country boys and girls in the high school and they managed it by living with relatives in town or by boarding with a family during the week. They might go home on weekends, but that was not always a sure thing because of the nature of the roads. In wet weather in northern Missouri one was more apt than not to be immobilized by deep mud.

During the last year or so of grade school and at the beginning of my high school years I had become enamored with the life of the outdoorsman. To me there was high excitement in the contemplation of life in the wilds, in the idea of tramping through forests or across mountains with a pack on my back, in the thought of sleeping out beneath the stars. It was an impossible dream for one of my age and one situated as I was in the heart of the Midwest. The only forests with which I was acquainted were narrow strips of deciduous woodlands along the muddy streams that cut through the cornfields of northwestern Missouri, and certainly there were no mountains, even though there were low ridges that formed the borders of broad river and stream valleys. All in all it was an unexciting landscape for an embryo adventurer.

To go back a few years, I had been greatly stimulated by the writings of Ernest Thompson Seton, who was truly a demigod to me. I read his animal stories over and over and knew them almost by heart. But for me the real gem of Setoniana was his classic *Two Little Savages.* I discovered that book in our shelves—it had belonged to Phil—and once having found it I devoted myself to it by the hour, month in and month out. It seemed as if I never tired of it. It was a magic book.

Seton knew exactly how to set forth for boys (and for girls, too) the intricacies and joys of woodcraft, and his text was made memorable by the numerous crisp drawings that he placed in the margins of the pages. He knew what he was doing; he was a good, all-around naturalist and an imaginative artist. *Two Little Savages* was my bible for several years and I thought that E. T. Seton could do no wrong. (Years later I had the pleasure of meeting Seton on several occasions when he came to the American Museum of Natural History in New York, and once at his "College of Indian Wisdom" outside Santa Fe, New Mexico.)

There was another magic book in my life at that time—a little pocket bird guide written by Chester A. Reed and illustrated with color plates. Each page contained a picture of one species, with a short descriptive text accompanying the illustration. The book had a soft but durable cover, and it would fit easily into an ordinary suit or shirt pocket. It went with me into the woods and fields during several years of my growing up, and it did help me to become acquainted with a considerable array of local birds.

With the Seton book in my memory and the bird guide in my hand, I frequented the woods, such as they were, and the fields that formed the rural setting surrounding my hometown. Most of my excursions were one-day jaunts, or even more commonly half-day trips, generally made in company with one or more of my friends, specifically Paul Diss and the two Green boys, Harold and Eldon. It was Harold who introduced me to the little pocket bird guide that was so much of my life for several years.

My first real camping trip was with the Boy Scouts, and that was in the summer of 1918. There were two more camps during the following two summers, and then the Boy Scout movement in Maryville seemed to collapse. But the three camps, in 1918, 1919, and 1920, were memorable experiences in my young life.

Paul, Harold, Eldon, and I were not totally dependent on the scouting movement for camping and other outdoor pursuits. We had our own organization, an Indian tribe christened the "Little Bears," and many a day we explored the meager woods of northwestern Missouri. We even had our own camping trip one summer, when we set up a tent and roamed the wooded borders of White Cloud Creek, which flowed through a farm belonging to an uncle of the Green boys.

In the summer of 1921 (following my sophomore year in high school) it became apparent that there would be no Scout camp. A year or so earlier the Greens had left town and moved to Hannibal, Missouri. One day Mr. Green called our house and asked if I could come to such and such a house in town, where he was visiting relatives. He also called the Diss household.

Paul and I went to see him, and were immediately more than pleased when he proposed that we come to Hannibal that summer to visit the Green boys and to go camping with them. Arrangements were made, including, of course, consent from our parents. Then came the period of impatient waiting until the day arrived for us to depart for Hannibal.

I don't suppose the time involved was particularly long but it seemed all too long for me. Every morning as I worked in the big garden behind our house, I would stop my hoeing or whatever I might be doing when the local train came coasting down the long hill behind us on its way southeast. For several days I watched it carefully because I had learned that there were two trains, running on alternate days, one being a very prosaic affair with a couple of standard coaches and a baggage car, but the other having as its rear coach a moderately magnificent creation with a back platform and large rear observation windows. There were two final seats on each side of the aisle in this coach facing backward, so that their occupants could sit in splendor and watch the receding track forever growing progressively narrower and finally coming to a perspective point in the distance. I was hoping that we would be lucky enough to ride on that train, and furthermore lucky enough to occupy one of the two rear seats.

On the day that we boarded the train the splendid observation car was

Our local train was a very prosaic affair with a couple of standard coaches and a baggage car.

on, and joy of all joys, the rear seats were unoccupied. (Perhaps the passengers that were on the coach when we got on did not like to ride backwards.) Paul and I established ourselves in one of those seats and got ready to enjoy our journey, watching the scenery retreat as we rattled along at perhaps thirty-five miles an hour. When lunchtime came we ate some sandwiches and other comestibles that our mothers had prepared for us, and then we sat quietly satisfied through the afternoon as the train continued on its way.

Late in the afternoon we arrived in Moberly, Missouri, where we were destined to change to another train for Hannibal. It wasn't a long ride, and we arrived in Hannibal about ten o'clock at night, where we were met by the entire Green family. We were tired; soon we were in bed and oblivious to the world around us.

The next morning we went downtown with the Green boys. First of all we visited Mr. Green, who was proprietor of a drugstore. Not far away the great Mississippi rolled its muddy way south beneath a bright morning sun. As we were hanging around in the back of Mr. Green's drugstore, doing nothing in particular, we heard a loud whistle blast, and in an instant the Green boys were galvanized into activity. "Steamboat!" they shouted and out the back door they went, with Paul and me right after them.

We all ran to the levee, a broad brick pavement that sloped down to the river to disappear beneath the muddy wavelets. As we approached the levee we saw out in the river a sternwheel steamboat approaching the landing. Its two tall funnels were belching black smoke, the long gangplank, supported by a sort of crane on the front of the boat, was extended out ready to touch the shore, and on the lower deck at the front of the steamboat was a gang of black roustabouts, laughing and indulging in some mild horseplay. On the upper decks some people stood by the rail, and behind them were the doors of the cabins, extending the length of the ship and facing the narrow deck.

The steamboat approached the levee, its stern paddlewheel churning up a froth of water, and then I saw the name *Belle of Calhoun.* When the boat came to within a few feet of the edge of the water the long gangplank came down, and then things began to happen. The roustabouts started coming off the ship, each man carrying an object. As each man reached his destination, which was a sort of warehouse on the shore, he deposited whatever he was carrying and picked up something to take back to the ship. Thus there was a double line of men, coming and going, and carrying things off and on the steamboat. Almost always each man carried one parcel, no matter what its size might be. (Of course very large packages required the efforts of two or more stevedores.) Generally speaking, however, it was one man, one package, whether the object was a fifty-pound crate or a little box six inches square. This went on for some time, while we sat on some barrels at one side and watched the proceedings.

Finally the loading, off and on, was completed. There was a blast on the whistle, the gangplank went up and the sternwheel began to rotate, to back the ship out into the stream. Soon the boat was squared away, the sternwheel reversed direction, and the *Belle of Calhoun* began its trip downriver. And as it went on its way I saw one of the roustabouts dip a bucket on a rope into the river and pull it out full of muddy water. Then he and his companions gathered around to drink the water, muddy as it was. That was my last sight of the riverboat, as it churned its way down the broad river.

It was a sight out of the past, almost. It was Mark Twain come to life. It was something that probably has not been witnessed now for many years, and I was indeed fortunate to have seen it.

Thus began my weeks in Hannibal and environs. Paul and I went with some Hannibal Boy Scouts on a little summer camp on the banks of Salt Creek, a stream that debouched into the Mississippi. The camp, however, was only a part of our summer. We spent many happy days in the countryside around Hannibal, modern-day Tom Sawyers, hiking and swimming and exploring. We spent hours along the high bluffs that bordered the river, wandering through the woods and picking wild blackberries in open, sunny fields. We crossed the river on the railroad bridge, to swim in clear bayous on the Illinois side. We went underground, to explore the "Mark Twain Cave," at that time an undeveloped limestone cavern of great dimensions, the locale for some of the adventures of Tom Sawyer. There were no electric lights installed in the cave, as I think now is the case, so we made our way through the passages each of us holding a candle in his hand. Luckily the Green boys knew their way around so we didn't get lost. It was an adventure, no doubt about it. So the weeks passed, and finally it was time for me to return home.

It was at about this time that I struck up a friendship with Marvin Westfall, a companionship that was to last until we grew to the estate of young manhood and went our separate ways. One day I was walking along Sixth Street, a couple of blocks from our house, and there in an empty lot next to the Westfall house was a small teepee. I promptly did some investigating, and I found that it had been set up by Marvin, who until then had been an acquaintance, but not a particularly close friend. It soon turned out that Marve, like me, was a fan, one might almost say a worshipper, of Ernest Thompson Seton. So the two of us formed a woodcraft and hiking association (if such a formal word can be used for two like-minded boys) and for several years we hiked through the countryside around Maryville and explored the narrow woods that bordered the rivers and streams. It was a healthy sort of activity, and I suppose we learned a bit about the natural history of northern Missouri. We even found some shale banks along a little stream known as Florida Creek, where we collected fossils of Carboniferous age. Between times we spent many hours at each other's homes.

Marve Westfall (left) and I would go out into the woods near Maryville.

My interest in nature and the out-of-doors continued, from the beginnings in grade school through my high school days. This interest was to have a strong bearing on my adult life; because of it I eventually became a paleontologist, destined to spend my years in museums and on field expeditions to all of the continents of the world.

It was in the summer of 1922, following my junior year in high school, that I learned to drive. The whole procedure was very informal; there were no driving schools, no examinations, and no driving licenses to bother with. My brother Phil, home for a few days for a vacation, got me started. He took me out in the country in our Dodge touring car and tutored me in the essentials of getting it over the road. After that I was on my own, and within a few weeks I was a pretty competent driver. Of course I didn't face all of the driving pressures that exist today; traffic was very light, extremely light on the dirt roads in the country, and speeds were slow. Thus by the end of July I was able to handle the car with a good deal of skill, which was lucky, for early in August we were to start on a trip to Colorado.

Plans and preparations were made, and finally we were off. The old Dodge had a cloth top, as did all touring cars of those days. If the weather got bad, we put up side curtains with little isinglass windows (this was before the days of transparent plastics) which cut down visibility almost to the point of lateral blindness. Touring cars seldom had trunks at the back—that was where the *two* spare tires were carried. So we piled part of our duffel into the back seat on the left side. There was a reason for the left side, because on the port running board was clamped a metal lazy-tongs arrangement, which stretched from the front fender to the back. This served to make a sort of fenced-in space where additional luggage could be stowed, with a canvas cover tied down over it to protect the suitcases and other parcels from the weather. Such a system made it impossible to use the doors on that side of the car, and we were restricted to getting in and out on the right side. Which was the safe side anyway.

The first leg of our journey was from Maryville to Lincoln, Nebraska, a distance of about 120 miles. Halfway along we had our first blowout. In those days of high-pressure tires (70 pounds) blowouts were frequent. As I recall, we had three of them in the 600-mile journey from Maryville to Denver. That's why we had two spares on the back of the car.

We had several pleasant days in Lincoln with Phil, who had just started as an engineer with the state highway department. Then we were on our way, across the length of Nebraska and half of Colorado on dusty dirt and gravel roads. We bowled along at the comfortable speed of twenty miles an hour; on occasion when the roads were in good shape we would reach a breathtaking rate of twenty-five miles an hour. It was a three-day trip; one day to Oxford, Nebraska, one day to Sterling, Colorado, and the third day into Denver.

It was on that third day that I saw before me the marvelous sight of the Colorado Front Range looming up ahead of us, a blue mirage tipped with white, floating, so it seemed, in the western sky and merging with the clouds. It fairly took my breath away. As we drove ever toward the west, and the mountains became increasingly distinct, I found myself more and more entranced by and attracted to them. Those great mountains were for me a new golden land.

In Denver we visited relatives, after which we set out for our ultimate goal, a hostelry called Moraine Lodge situated in Moraine Park, five miles or so to the west of Estes Park village. We drove north through Longmont to Loveland and then turned west to enter the narrow confines of the Big Thompson Canyon. In those days the road was a winding, narrow dirt affair that skirted the edge of Big Thompson Creek all the way. This led us to the village and then we went on from there to the lodge.

Moraine Lodge consisted of a central log structure, housing a dining room and a kitchen, a lounge, and I think a few bedrooms. Scattered around it were some cabins and a number of tents with wooden floors. We had one of the tents. It was delightfully primitive in a way. Every morning I would be awakened by chipmunks running up and down the roof of our tent. Then it was out of bed for me, and a cold wash-up at a basin supplied by water from a pitcher standing on the washstand. Outside we had a magnificent view across the park and beyond to Long's Peak, one of the high points of the Front Range, looming up to more than 14,000 feet, its eastern face formed by a sheer cliff of terrifying proportions. (Today Moraine Lodge is no longer an inn; it has been taken over by the Park Service and serves as a visitor center.)

Those two weeks in the Rockies were a magic time for me. They opened new perspectives that changed my thinking in a profound manner. I fell in love with the lofty mountains, with the clear, blue skies, with the sparkling streams and lakes, and with the general feeling of renewed vigor that comes from breathing pine-scented air at high altitudes. I wanted to stay forever; I was more than reluctant to return to the prosaic cornfields and the heavy atmosphere of the Missouri Valley.

One of the cabins at Moraine Lodge was inhabited by a dignified old gentleman and his wife. They were regular summer residents, who had been coming to the lodge for quite a number of years. He was none other than Thomas Watson, the man who, as a young technician, had worked with Alexander Graham Bell on the invention of the telephone. It was Watson to whom the first words ever spoken on a telephone had been addressed. Bell and Watson were in separate rooms on the top floor of a building on Exeter Place in Boston, where they were trying to get their system to function. Suddenly Bell, who was having some problems with the apparatus, involuntarily said, "Mr. Watson, please come here! I want you!" And at just that moment the proper hookups, or whatever they were, had been achieved, so that Watson heard Bell's voice over the instrument. It was a moment of high jubilation for both men.

The Watsons would invite us to their little cabin and Mr. Watson would tell us about his work with Bell, or more often than not would read us a story. He loved to read to people. He was very nice to me and would invite me in by myself to tell me about his experiences. Indeed, he inscribed for me a little booklet he had written about the invention of the telephone. I enjoyed those sessions and have always remembered them. Here, as in the case of my Grandfather Adamson, an individual lifetime spanned almost inconceivable events in the history of modern man. With my Grandfather it was the time from the unsettled West and the unbroken plains to the time of the airplane and radio. With Mr. Watson it was the time from the very concept of the telephone to the present highly sophisticated development of this means of communication.

All too soon it was time for us to pack up and return to Missouri, a trip I did not in the least wish to make. I wanted to stay in the mountains for the rest of my life, but that wish was something less than realistic. I drove my parents down the St. Vrain Canyon and out on to the plains, and we set our course eastward for the drive home. All the way I had visions of the mountains in my head, and they stayed with me for a long time afterward—indeed through my entire senior year in high school. I graduated in the spring of 1923, with ideas of perhaps going into the profession of forestry.

That summer was my summer of discontent; I would dearly have loved to go back to the mountains, but since that seemed out of the question I wanted some kind of job. Unfortunately I was at that in-between stage, of being not quite far enough along for a real job but too far along to be satisfied with the summer chores around our home. So for much of the summer I was bored stiff.

In August, however, we drove north with friends, the Corwins. The destination toward which we were headed was a log cabin on a lake in Superior National Forest, just short of the international boundary line. We drove to Grand Marais, Minnesota, and then some forty miles on a very primitive road to its end, where we left the car. Then we went by canoe across Hungry Jack Lake, took a portage to Clearwater Lake, and by canoe again to the camp where we would be staying. We had a pleasant enough time; I made some little day-long canoe trips here and there and explored the shores of Clearwater Lake. It was wild country all right, but I felt hemmed in by the forested flatness of the great Canadian Shield. I still longed for the mountains.

In the course of things we went on into Ontario and on the way back passed through Chicago, where we stayed for several days. Here I visited the Field Museum, which had only recently moved into the vast classical-style building that it occupies on the Chicago lakefront. It was a wonder house to me, and I suddenly had the idea that work in a museum might be a very good way to spend one's time. I even mentioned this to my parents but since the idea did not seem very practical just then, I returned to thoughts of forestry.

Although I had forestry in mind, I really did not feel certain that I wanted to make that my lifework. So that fall I entered the college in Maryville. The school had advanced from its Normal School status, to become a State Teachers College, and while it was primarily oriented to the study of education, it had on the staff some excellent teachers of science and of certain courses in the humanities. Consequently, I felt I would not be losing any time if I stayed in Maryville for a year or two, until I could establish the direction I wished to follow. (As it turned out, I stayed there three years, during the course of which I received some good undergraduate instruction.)

During the winter of that first year at the college in Maryville I wrote letters to several schools of forestry, asking for their catalogues. In addition I wrote some letters to the superintendents of several National Forests in the Rocky Mountain region, applying for a summer job. There were some rejections, but to my pleasant surprise I received one day a letter from J. V. Leighou, Supervisor of the Arapaho National Forest in Colorado, offering me a summer job working on trails in that forest. Of course I was happy and excited at the prospect; now I could return to the mountains and get some real experience in forestry.

3. ROCKY MOUNTAIN DAYS

After I received that letter from the Supervisor of Arapaho National Forest the spring days of 1924 were a time of anxious anticipation for me. Anxious because the time for my scheduled departure was all too slow in coming, and anxious because I did not know just what was ahead of me. Day after day I waited, hoping for a letter from Supervisor Leighou, informing me when and where I should report, but the letter never came. Finally in desperation I wrote to him, and shortly thereafter I did get the promised letter. It seems that he had been very busy, and a letter on such a minor subject as one more temporary summer job had become sidetracked somewhere in the office. It was a cordial letter and told me when and where I should make my appearance: as soon as convenient, in Hot Sulphur Springs, Colorado.

Hot Sulphur Springs was a drab little town of a few hundred residents, with the Forest Service offices on the second floor of the bank building. There I reported and there I was assigned to work for John Glendenning, one of the District Rangers.

For a few days I helped him build a fence near his house on the edge of town, to contain his horses, and then we made ready to leave for my work area. For some reason we left late in the day, riding horses and leading a pack horse. We rode through the evening twilight and after that for several hours amidst deep forests made eerie by bright moonlight filtering through the branches of the trees. We camped at last, and the next morning we started the day by hunting for the horses, a procedure that I soon learned was rather standard practice.

We rode on, and late in the afternoon we came to a little pocket in the hills known as "Lost Gulch," and there, while a mountain rainstorm lashed us, we unsaddled and unpacked and got ready to make camp. The camp was a simple affair, an A-tent set at the edge of the trees, with a pretty little mountain meadow traversed by a stream out in front. We had our evening meal and then turned in.

Bright and early the next morning we were up, and after breakfast Glendenning left me all alone in Lost Gulch. Needless to say, he left most of the food that we had packed in and some other supplies, but there I was as he rode away, on my own, isolated, with no horse and the nearest person I don't know how many miles distant. Glendenning promised to return in two or three weeks.

My assignment, believe it or not, was to wander around the mountain meadows in my vicinity, digging up larkspur plants with a mattock. Larkspur is poisonous to cattle. The cattlemen evidently had put pressure on the Forest Service, so the government boys were trying an experiment of manually eradicating larkspur in places where cattle might graze. And I had been elected to be the experimenter.

It was a hopeless and boring task. All day I wandered about looking for larkspur plants among the sagebrush and mountain grasses and whacking out the plants wherever discovered. Those days, all by myself, were unbelievably long and tedious, and I was always glad to get back to my tent,

Hot Sulphur Springs was a drab little town.

where I could fix my evening meal and lose myself in a night of sleep.

Yet in spite of the boredom and the recognized uselessness of my assignment, there were some compensating joys. One of them was listening to the veeries sing their evening songs. I had never before heard the western veery, a cousin of the hermit thrush, and I shall never forget the pure, descending tones of their song—like glass marbles rolling down a glass spiral. The songs echoed back and forth through the pine trees as the light faded and the air grew chill.

I managed to feed myself sufficiently well and was able to supplement bacon and beans with trout that I caught in the little stream in front of my tent, trout caught in a very primitive fashion, with a willow pole, a piece of line, and a grasshopper on the hook. Perhaps it wasn't very sporting but it did provide some fresh protein.

So I lived, and so I was living when Glendenning appeared one day at the end of about three weeks and informed me that I was being shifted, to work with a crew building trails. Whereupon we packed my gear, including the detested mattock, and set off for the camp of the trail crew only a few miles distant.

That trail crew was carrying on a big operation for those times and in that place, the biggest I ever was involved in during my Forest Service days. Even so it was no great shakes. There were the boss of the outfit, a burly and rather hard-nosed character, and his wife, who served as camp cook. Their little boy was along, too. Then there was a bearded giant, Harry Douglas, who became my particular friend. Finally there was an old Swede, Ben Tandall. We had a string of a dozen or so horses, including a matched team of heavy sorrels that belonged to the crew boss.

We were building new trails through the trees and across mountain slopes. Trails were especially needed in those days for access to the forests—for patrols and for fire fighting. That was long before the days of bulldozers, helicopters, and fire-fighting airplanes. Ours was the job of establishing trails where hitherto there had been no trails.

If we were working in dense forests we would cut down trees where the trail was to be, making a path through the woods wide enough to accommodate a fully packed horse, say about six feet or so in width. Where the graded trail was to be we would cut the trees as high as we could reach effectively with an ax, say five or six feet above the ground. Then the boss would hitch a heavy chain high on the long stump, attaching the other end to the double-tree of his team. The horses would lunge forward, and with the leverage of the high stump the tree generally could be pulled out by the roots.

All of this was good training for me, and through the hard school of experience I learned how to use a double-bitted ax, swinging it either from the right or the left side, and I learned how to manage my end of a long, two-man saw, for those were the days before gasoline-powered wood saws. Also I learned how to use a double-bladed plow with iron handles.

After we had swamped out the path for the trail we would plow a deep furrow with the heavy plow. Then the horses would be hitched to a little triangular drag known as a Martin Ditcher, and by pulling this along the furrow would be converted from a ditch of sorts into a flattened trail. Such was the procedure when all went well.

More often than not things did not go so well. In the deep woods where there was a thick forest floor it was not so bad, but on the exposed slopes the job of plowing and grading was anything but easy. We would generally be going across a steep slope, so the horses were positioned with one much higher than the other. This made it hard for them, and their problems were not lessened by the necessity for keeping their footing on an uneven, rocky slope. The boss would take the reins and urge the horses forward with loud, blistering shouts. The horses would make a lunge, trying to keep their footing while at the same time pulling the plow. And there I was at the handles of the plow, trying to hold it on course. About every five feet the plow would strike a large underground boulder, whereupon it would fly up in the air, and I would likewise sail in a wide, parabolic arc, landing twenty feet or so down the slope. Then I would laboriously climb back, and we would position the plow and try again. Perhaps it is now evident why the plow had iron handles; wooden handles would have been splintered into kindling before half a day had gone by.

It was all hard work, holding the plow or clearing the path with ax and saw. But I thrived on it in most respects. In one way I didn't have it very good, because the boss was a hard-driving martinet and he took out his frustrations on me since I was the greenhorn of the crew. That was where my friendship with Harry Douglas paid off. When things got too bad Harry came to my defense, and Harry was the sort of large, impressive woodsman with whom no one would want to tangle. Also Ben Tandall was a friend.

Then one day things improved for me. We were swamping out the trail, and I grabbed one end of a fairly large log, perhaps eight feet in length, to swing it out to the edge of the trail. The boss was bending over to pick up another log. My log was heavy, it got away from me, and came down with a sickening thud on the back of the boss's head. He went out like a light and for several minutes was as cold as a fish. I was scared—no doubt about that. With some cold water from a canteen and some ministrations the boss came to, much to my relief. But he had a sore head and neck for a couple of days. He obviously thought that I had crowned him on purpose, because from then on he was just as nice as could be to me.

That work came to an end, the trail crew broke up, and I was shifted to another crew working under the direction of Ranger John F. Johnston in the depths of Arapaho Canyon. I made my way to my new assignment, and there at an old trapper's cabin on the edge of Arapaho Creek, a tumbling mountain stream, I joined my new companions, a young Swedish-American about my age named Milt Johnson and an old Swede named John Lind.

We slept in a tent pitched by the crumbling cabin (that ancient log structure was dark and damp inside) and cooked our meals at a campfire, next to which was a big hole where we would bury the Dutch oven when we baked biscuits, which was every day. We had no horses—indeed there was no place in the narrow bottom of that 2,000-foot-deep canyon where horses could be grazed—and we did all of our work by hand. We cut trees, and we graded trail with mattocks, and where there were rocks we blasted them out with dynamite. It was a confined way of life for the sun would not reach us until about ten in the morning and it would disappear over the western rim of the canyon about four in the afternoon. It rained a lot and the nights were cold. So passed the remainder of my first summer in the Forest Service.

I was not sure whether one summer in the Forest Service justified another, especially since I had received a letter during that next winter from Johnny Johnston in which he said that he didn't think there would be money for extra summer help. I decided to go to Colorado anyway on the chance that I might find work of some kind, and one morning I turned up at the Idlewild Ranger Station, which was Johnny Johnston's headquarters, to be greeted with welcome warmth by Johnny. He said he was just thinking of me and wishing he could get in touch, because some unexpected funds had been appropriated and recently received. My luck was in—I was all set for the summer.

I didn't have to wait; we got right to work and packed up for a trip to Arapaho Canyon, where I had left off the previous summer. I was to be the leader of a two-man crew, the other member being a local boy whose name I have now forgotten. It was a Czech name. Anyway, he was called "Dutch."

The next day Dutch and I were on our way, and for the first and only time I got lost in the mountains. I wasn't exactly lost for I knew approximately where we were, but I couldn't find the trail. It all happened while we were crossing a huge mountain meadow. We emerged from the forest on one side, but in crossing the meadow, which was all grass with no trail and no trail markings, we failed to take the proper direction, and so we missed the trail on the other side. We did a lot of hunting back and forth with no success. Finally, since the afternoon was beginning to pass us by I decided that we would cut through the forest toward the west, for I knew that if we did that we would eventually strike Arapaho Canyon. So we did.

It was a tedious journey, breaking our way through the dense forest, and especially getting the pack horse around tangles of fallen timber. However we made it and arrived at the rim of the canyon an hour or two before sunset. The next task was to get ourselves and the pack horse from the top to the bottom of the canyon. Somehow we managed it, although the going was so steep that the pack slipped right over the horse's head on four separate occasions, and each time I had to repack him on the steep slope. We reached the bottom about dark and made camp.

When Johnny Johnston came around later on an inspection trip he shook his head and remarked with wonder that he didn't see how in the hell we ever got that horse down the canyon. Indeed he brought the subject up several times during the summer.

Dutch did not stay with me long. He was a boy of the mountain towns, and he didn't especially like the forest. That left me at loose ends, so Johnny Johnston decided that I should team up with Sam Stone, an old mountain man type who had just become available for trail work.

Sam and I joined forces, to work on trails in Buchanan Canyon, which joined Arapaho Canyon just above Monarch Lake, a long, narrow body of water. There was an old logging camp of sorts at the head of the lake where the two canyons came together, and that was to be our staging area. The camp consisted of several decrepit log cabins at the end of a very rough road that ran along one side of the lake. On this road old Sam could just make it to the cabins with his car, a Model T Ford with a Ruxtell Axle. The Ruxtell Axle was an accessory that could be added to the Model T, converting it into a sort of primitive jeep.

That summer was the time when we did quite a lot of blasting. We were working in rocky country and the trails we were building or repairing often were blocked by big rocks that had to be removed. We therefore resorted to dynamite, and old Sam was the dynamiter. He had a lot of experience along this line, which was one reason he had been chosen to work in this particular area. Thus the days went by, with us cutting trees and grading trail and frequently setting off blasts that made the mountain welkin reverberate from peak to peak.

One day Sam and I went into the nearest town—Granby it was—for supplies. Then we drove back to the camp and packed the horses for a foray up Buchanan Canyon to a place beyond where we had been working. We established our camp and began our work, and within a day or so Sam informed me that he was going back into town; he was going into town to get a lady, and that statement made me raise my eyebrows. Whereupon Sam informed me with some heat that this was all very proper; the lady was a medium, and he was going to bring her up to locate the "lost mine" in Hell Hole, a section of howling wilderness far removed from our camp or from any place else for that matter. Sam proposed that I join forces with him; he would pay me as much as I was getting from the government, and perhaps we would strike it rich. I declined with thanks.

There are more lost mines in the mountains of western North America than is readily imaginable. But their origins are monotonously similar. Some old prospector or a pair of prospectors had, according to the often repeated story, located a mine of superb wealth, but then this worthy or the partners had died before he or they could inform anyone else as to the exact location of the mine. There it was, waiting to be rediscovered.

It should be added that the lost mine in Hell Hole had been originally discovered by a pair of Swedes. Old Sam had visited the medium in town, and she had been in astral communication with the two Swedes. Of course their spirits had appeared, to tell the medium how to find the mine.

Sam deserted me, taking the horses and quite a store of food, not to mention a fifty-pound box of government dynamite. I went on working on the trail as best I could, and two or three days later Sam came back up the trail with a pack horse in tow and the medium following most uncomfortably on another horse. Poor medium! She was a little old lady, seventy-five years old at least, and not at all suited for a rugged pack trip into the wilds. They stopped and chatted with me for a few moments and then went on their way. And that was the last I ever saw of Sam Stone.

I continued my work, an isolated trail stiff in the high mountains. My camp was by a stream in the middle of a mountain meadow, a location beautiful beyond description. Behind me the peaks rose in majestic and at times forbidding grandeur to the crest of the Continental Divide. In front of me was the grassy meadow, replete with alpine flowers. It was a setting that made my forced exile quite tolerable.

I knew that Johnny Johnston would be along sooner or later on an inspection trip, but in the meantime I was marooned, after a fashion, with a comfortable camp and plenty of food, but with no horses to get me out of the place. As I worked day after day I began to have hallucinations. I would hear voices, I would hear church bells ringing, and above all I would hear the footfalls of horses on the trail. It is quite an experience to be absolutely alone, an experience that I suppose doesn't come to many people these days.

Then one evening at suppertime I went down to the little stream for a pail of water. As I turned around, there were Johnny Johnston and the Forest Supervisor, Jack Leighou, sitting on their horses and grinning at me like a couple of Cheshire cats. The tumbling waters of the stream had masked the sound of their approach.

Well, they asked, where was old Sam Stone? I told them. And then they didn't know whether to be annoyed by Sam's defection and especially by his appropriation of government property, or whether to be amused by the absolute ludicrousness of the situation. They made camp with me that night, and around the campfire they were more amused than annoyed. Indeed, we had a hilarious evening, and they gave me a good deal of chaffing as to whether I would be the next one to go off his rocker. At any rate, it was decided to move me back to Arapaho Canyon, to work with Milt Johnson, my companion of the previous year, and another trail worker, Clarence Murphy.

The next day we broke camp and made our way back down Buchanan Canyon. As we started up Arapaho Canyon we met Milt Johnson coming down. Where was he going, my Forest Service bosses inquired. It seems that Milt had become very fond of a girl living in one of the Middle Park villages, so he was quitting the Forest Service, to take a job where he could be near the object of his affections. He went his way down the canyon and we went our way up, with Leighou and Johnston making more merry quips about loony trail stiffs and asking if I were to be the next on the list.

We arrived at the trail camp, where we found Clarence Murphy wondering just what he was to do. I joined Clarence and we spent the rest of the summer together in Arapaho Canyon. We immediately hit it off and had a most pleasant time working with each other. Clarence was a veteran of World War I (in those days such veterans were still young men) and many evenings as we sat around our campfire he told me about his experiences in France. He had been through some very heavy fighting, and for the first time I really learned from someone who had been there about the horrors of modern warfare.

Clarence and I made trail in that deep canyon, and we too did a deal of blasting. Clarence had a big dog, a Scottish deerhound named Speed. Speed had quickly learned what blasting was all about, and he didn't like it. When we would yell "Fire!" and run down the trail Speed would join us and crowd as close as he could to one of us as we took shelter behind our respective trees. He would stand there and tremble violently in expectation of the blast. Then when it went off he would be transfigured with joy and relief, leaping and barking with all of the energy at his command.

My summer ended, and again it was time to leave the forest and go back home.

I was to have two more summers in the Forest Service, both in Johnny Johnston's district. But they were to be different from my previous summers because in the meantime I had made a choice, as will presently be told, that would lead me away from forestry and into a very different field of endeavor—paleontology. Consequently, my final two summers in the forest would be a matter of working at something I liked but something I now felt would not be a prelude to my future. Perhaps that made my last two summers in the high mountains seem less serious in intent than the preceding seasons.

I reported to Johnny Johnston early in the summer of 1926 and was immediately teamed up with Jerry Yetter, a forestry student from Fort

Jerry and I ended that summer above timberline on the shoulder of James Peak.

Collins, Colorado, and a fellow about my own age. We had a nice summer together, working for the most part in areas that were new to me.

Jerry and I ended that summer above timberline on the shoulder of James Peak, where the winds blew and the horses stood with their heads down and tails to the weather. That was during the day. At night we rode down to a more clement environment.

Then came my fourth and final year in the forest. I began that season with an old coot named Pete Mills, and for a while we enjoyed the luxury of living in an unused ranger station on Byers Creek. It was a very comfortable log building, and at night I could lie in my sleeping bag on a cot, not on the ground, and read by the light of a kerosene lamp. I had an *Atlantic Monthly* along, and there I read the first installment of *Jalna* by Mazo de la Roche. For me that was a delightful literary adventure.

After about a month with Pete Mills I was transferred and worked the rest of the summer with Barney Dean, a college student from Iowa. We spent most of our time in Cascade Canyon, a picturesque, rocky place with one waterfall after another. We worked on trails in that canyon up to a lovely lake at the base of the Continental Divide. It was some of the finest mountain country I ever worked in. We also worked in my old stamping ground, Arapaho Canyon, and carried along to Arapaho Pass, where we camped amid the clouds.

The summer passed, and it was time to leave. Barney had a stripped-down Model T Ford, which he had parked at the foot of Monarch Lake, and in that primitive vehicle we drove over the Divide and down to Denver. My days with the Forest Service had ended.

Each summer, on my way from Maryville to Colorado to work in the Forest Service, and on my way back, I would stop in Lincoln, Nebraska, to visit my brother Phil and his wife True. Frequently when in Lincoln I would visit the university museum, where there was a considerable collection of fossil mammals from the Upper Cenozoic beds of the Great Plains. To me there was magic in those skeletons and skulls, in the forms of the bones, and in their textures and colors.

During the winter following my second stint in the Forest Service I got to thinking that perhaps fossils might be more interesting than trees. Having thought about this at length, I wrote to Phil in the spring of 1926 to ask him if he could find out what might be the possibilities of a career in vertebrate paleontology. It was a long shot on my part, and I truly did not expect much from it. I supposed that there would be a discouraging reply to my feeler, not on the part of Phil of course, but rather from the people at the museum. Somehow it seemed to me that a museum was a place beyond the reach of most mortals.

We spent most of our time in Cascade Canyon.

Imagine my surprise and joy when I got a letter back from Phil saying that he had talked with Professor E. H. Barbour, the director of the museum, and that Professor Barbour would like to meet me. That seemed almost too good to be true, but it was true. So plans were made.

For some reason, I don't remember why, my parents were to be out of town when I planned to go to Lincoln. I think perhaps they were on a visit to Herschel and his family. At any rate I was alone in the house, and I had to rise up at an ungodly hour to catch the early morning Wabash train, coming through from St. Louis to Omaha. I accordingly arranged to have Marve Westfall come and spend the night with me, for I figured that between us we might manage to get me up in time to make the train. As could be expected, such precautions were not necessary. I hardly slept all night, and well before train time I was up in a fever of excitement. Marve went with me to the station and helped get me embarked.

In Lincoln I was met by Phil, and in due course was ushered into Professor Barbour's office. We talked, and I showed him drawings of skeletons that I had made in biology classes. I told him what I had in mind, although what I had in mind was, I'm afraid, a bit vague. Nevertheless, it seemed to satisfy Professor Barbour, and the interview ended on a very satisfactory note. If I should come to the University of Nebraska the next autumn I would be given a position as a student assistant in the museum. Needless to say, I was in something of a state of euphoria.

After a weekend with Phil and True, I returned to Maryville. At last the course of my life had been set. I was to spend two more summers in the Forest Service, but now my efforts were to be directed toward making for myself a career in vertebrate paleontology, and as an adjunct to that a parallel career in museology.

The final days of the spring of 1926 were my last as a regular resident of the house on East Seventh Street. The time for a break with my early life had come. From now on I would journey to Maryville as a visitor. First I would go to Colorado for the summer, and after that I would return to my boyhood home for a brief visit. Then I would leave for the university and the beginnings of my new life. I had found my direction.

4.
AN APPRENTICESHIP

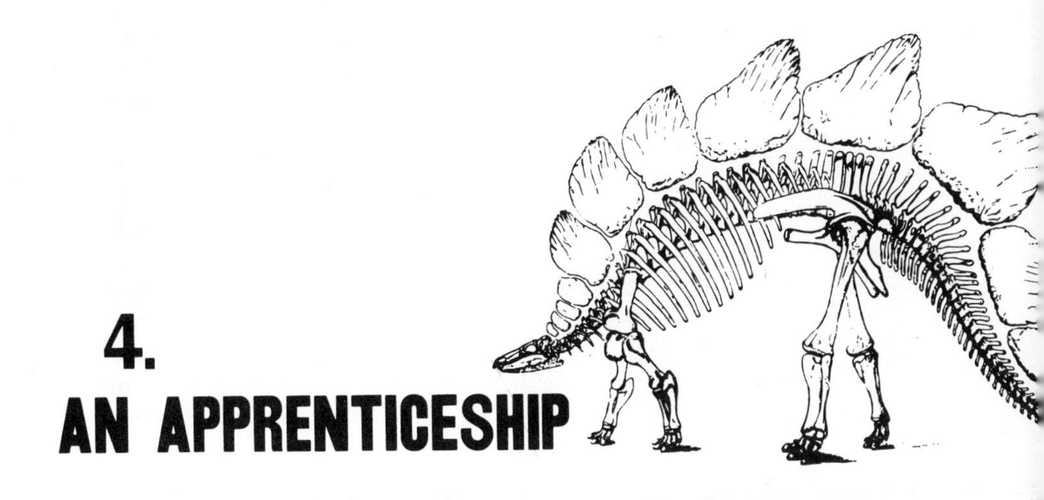

In the fall of 1926 I found myself at the University of Nebraska in Lincoln, an undergraduate with advanced standing, yet none the less terrified and homesick as the greenest of freshmen. The university seemed monstrously large and impersonal, indeed it seemed threatening to me. Lincoln, a city of modest population and dimensions, seemed like a veritable metropolis. Those first months in my new environment were trying and in some ways unhappy months. Perhaps, during that euphoric spring of 1926, when arrangements had been made for me to attend the university, I thought I had found myself, but now I seemed to be floundering in a mildly hostile world.

I had come to the university because of the fossil mammals that I had seen in the museum; some of them beautiful, such as the skulls and jaws of ancient horses—the mineralized bone creamy white and the teeth dark gray; some strange and exotic, such as the skull of *Syndyoceras*—a ruminant with four horns, two above the eyes and two on the front of the skull; some truly magnificent, such as the massive bones of mastodonts and mammoths. These fossils evoked visions of a world long vanished, when Nebraska was a land of lush savannahs inhabited by hosts of unfamiliar animals, a land innocent of the presence of man. But during my initial months in Lincoln I had little contact with the fossils. Much of this was owing to the fact that there were various lectures to attend, laboratories in which to work, and subjects to be studied. And it was here that I felt the threat of the institution. Some of my lecture courses were largely attended, and I felt keenly the lack of personal contact with the professor that had been my experience in a small college. My classroom and laboratory sessions seemed exercises in survival—a constant effort to keep up with my assignments, to keep from being weeded out and dropped. Undoubtedly much of this feeling was in my mind and was exacerbated by the newness of my surroundings. At any rate, I did survive.

It was in the museum that I found refuge from a demanding world. Here I could look at the fossils, even though at first I did not have very close contact with them. They were objects of frequent admiration, and I got to know some of them as old friends.

The university museum in those days was housed in a far from handsome, four-story brick building that began at a campus street and came to an abrupt end a little way back. It had been built, I suppose, around the turn of the century, with the intention of extending it at some future date; but the date had never arrived. There it was, facing the world foursquare in one direction and ignominiously truncated in the opposite direction. Ironically, the entrance was in this flat, supposedly temporary back wall of the structure.

It should be said that a new museum was being built at the time, on another part of the campus.

Presiding over the museum was my ultimate boss, Dr. Erwin Hinckley Barbour, the director. Dr. Barbour at that time was well past middle age, a tall and rather distinguished-looking man with gray hair, a sort of Imperial beard, and pince-nez spectacles with a gold chain looping back to a semicircular gold-wire guard over the right ear to keep them from escaping into space and descending to the floor. Especially when he was excited.

Dr. Barbour was a gentleman of the old school. He had been a student and an assistant of O. C. Marsh of Yale, a legendary figure in the annals of American paleontology. Barbour, like the other men who worked with and for Marsh, was not particularly fond of that pioneering pirate of the early days in the history of western fossil exploration and eastern research, but he stuck it out and got his doctorate at Yale.

As may be imagined, Dr. Barbour had some tales to tell about his student days at Yale. In those distant times Marsh was directing a large and ambitious program of collecting fossils in the still untamed West and of preparing and studying them in New Haven. For this he needed numerous assistants, of whom Barbour was one. At the same time Edward Drinker Cope of Philadelphia was also carrying on a program that paralleled the work of Marsh, to such a degree, in fact, that the activities of the two men overlapped and collided. Cope, a brilliant man of Quaker heritage, had his disciples, among them Henry Fairfield Osborn and William Berryman Scott, then students at Princeton and eventually to become paleontological authorities of world renown.

Marsh and Cope were initially friends, but very quickly they became implacable enemies. Both men had money, both were independent and self-willed, and North America, especially the land west of the Mississippi River, was not large enough for them. Consequently they carried on a sci-

Dr. Erwin Hinckley Barbour was a gentleman of the old school.

entific feud of unparalleled bitterness; in fact it became a very personal feud. Of course the men who worked for them became at times involved in the row, generally against their wills.

Dr. Barbour told how one time Osborn and Scott came to Yale to study some of the fossils that Marsh had collected. They were Cope men, and that was anathema to Marsh, yet Marsh could not deny them entry to a public institution. However, he instructed Barbour to cover up and hide all of the specimens that he did not want Osborn and Scott to see. Then during the whole time of their visit the famous Professor Marsh, director of the Yale Peabody Museum, skulked among the nooks and crannies of the building, his feet encased in carpet slippers so that he would not betray himself, and from around corners and behind storage cases spied upon the two young followers of Cope as they quite innocently studied the fossils that had been made available to them. It must have been an interesting performance, especially for Barbour as an onlooker.

This is hardly the place for Cope and Marsh stories; they are legion. But the above is a small sample, passed on to me by one who was peripherally involved in some of the scientific shenanigans of those two great rivals.

Not long after completing his studies at Yale, Barbour came to Nebraska, during the waning years of the nineteenth century, to found and build up the collection of fossils at the university, particularly the remains of Cenozoic mammals. The high plains of the western part of the state offered a rich field for his collecting activities, and he made good use of his opportunities. Thus he took his place with the early founders of vertebrate paleontology in North America.

He had his idiosyncrasies. (Parenthetically, it would seem that all those who get involved with fossil vertebrates have their quirks. It is a field of activity that attracts people who somehow get deflected from the usual paths of human endeavor.) Barbour was a kindly person, but he certainly was not thoroughly organized. I often wondered during my years in Lincoln whether he ever had the day mapped out in his mind when he entered the portals of the museum each morning. All of which made him delightfully unpredictable, and at times a frustrating person under whom to work. Nonetheless I was fond of him, for I felt that he had my well-being at heart.

Among other things, Barbour thought of himself as something of an artist, or at least as someone who understood art, and I suppose it was for this reason that he insisted I take some art courses, especially studies in perspective and modeling, when I came to the university. It seemed at the time like a strange departure for one who was aiming at the scientific study of extinct animals, but he maintained that it would give me some insight if I were to work in a museum and be involved with exhibits. Perhaps he was right.

Dr. Barbour was a great person for museum exhibits; he was not one to spurn this aspect of museum practice as is at times the case with scholars who work in university museums. In fact, there was a bit of Barnum in Barbour, for he liked to make a good thing better, and a large thing larger. In those days when the museum still occupied its old building space was very much at a premium. Among other things, the museum possessed a truly gigantic mammoth skeleton which had been collected in past years. There was no room in the building to mount the skeleton, but on one of the upper floors, at the head of the stairs, Dr. Barbour had caused the forelimbs and the associated vertebrae and ribs to be set up to form an arch beneath which all who ascended the stairs must pass. It was effective, of that there is no doubt. And it was only one of many devices he employed to impress upon the public, and eventually upon the state legislators, that the museum had rare objects deserving a new home.

At the beginning of my Nebraska stint I did not, however, work directly for Barbour. I was assigned as a student assistant to Frederick George Collins, the curator of the museum. Mr. Collins was a roly-poly little Englishman who had emigrated to the United States during his middle age, when his daughter married an American lawyer, a Rhodes Scholar,

Mr. Frederick George Collins, a thoroughly charming and knowledgeable person, introduced me to what a museum is all about.

who eventually became an attorney in Lincoln. For most of his life, Collins had been a haberdasher in Exeter, England, but during those years he was keenly interested in geology and fossils. He had been associated with the Exeter Museum and thereby gained a good background in museum practice. He was a thoroughly charming and knowledgeable person and he introduced me to what a museum is all about.

Of course I was completely inexperienced. My first task at the museum was to dustproof all of the exhibition cases housing the displays of minerals and rocks. This consisted in applying black passe-partout tape along the inside juncture of the glass door of each case with its wood frame, to form a seal of sorts. The tape had to be applied evenly, to make a nice-looking border, and that was tricky because the tape had to be moistened to make it stick. And to moisten the tape effectively it had to be *licked.* A moist sponge just would not do the trick. During many long hours I licked what seemed to be miles of passe-partout, and that was wearing to my tongue and unfriendly to my sense of taste.

In time, however, I graduated from this menial task and began to help Collins curate some of the collections. I well remember the day that he decided we should get rid of some mineral and rock specimens that had no data accompanying them. (Dr. Barbour was a regular jackdaw; he would never dispose of anything no matter how inconsequential it was.) Collins conceived the idea that the simplest and most effective manner of disposal would be to stand at a third-floor window and heave the unwanted specimens into a rock garden below. There would be no visible evidence of their departure from the museum, and perhaps the rock garden would be enhanced.

We began successfully enough, but in the middle of our operation a passerby on the street thought that we were throwing rocks at *him.* Whereupon he started to bless us out in a loud voice, threatening to come up and give us the once-over. Collins and I backed away from the window, hoping that the strenuous objections of the party below were not wafted through the open windows of Dr. Barbour's office. Evidently they were not.

Another occupant of the museum during that year was Phil Orr, somewhat my senior and a part-time student. He was part-time because the more interesting segments of each day to him (and to me) were spent in the preparation laboratory of the museum, where he worked on fossils. I was able to put in a little time on the ancient bones under Orr's supervision, especially as the academic year wound down toward its end.

Phil was as much of a character as were the rest of us who worked in the museum trade; he was the only person I have ever known who could whistle two notes simultaneously. It was truly an experience to listen to him whistling in thirds as he worked away on a fossil bone. He also tooted around town and countryside on a motorcycle with a sidecar attached, a combination that seems to be virtually extinct today. At the end of the school year Phil left the Nebraska scene; he installed his wife in the sidecar of the motorcycle and off they went to Chicago, where Phil had been given a position in the paleontology department of the Field Museum. He was there for quite a number of years, and eventually transferred to the Santa Barbara Museum, where he pursued a very important program centered around the association of early man in North America with now extinct animals, especially mammoths.

Still another denizen of the museum during that first year of my Nebraska experience was Bertrand Schultz, who remained in Lincoln eventually to become the director of the institution.

In the late spring of 1927 the museum was moved into its new quarters, Morrill Hall, a busy operation in which we were all involved. Day after day during my available time I drove a venerable Ford Model T truck loaded with specimens of various sorts from the old building to its successor.

Most of the moving, however, was carried out by a trucking company which had been hired for the occasion. Needless to say, the cargoes that were loaded on the trucks for the short journeys between the buildings were quite unlike anything the movers had handled before, all of which made the work very interesting for the men. For one man especially, Henry Reider, the museum specimens, particularly the fossil bones, had a fascination of more than a casual sort. To anticipate a bit, during the following year Henry turned up at the museum seeking a job in the fossil laboratory. He was taken on and he spent the rest of his life there, becoming unusually skilled in the difficult technique of mounting fossil skeletons. More of Henry later.

A summer interlude followed the hectic going back and forth, the emergency decisions as to where to stow things at least for the time being, and the general confusion attendant upon moving a museum. This was the time during which I spent a final season in the Forest Service in Colorado. Then I was back at the museum for a second year as a student assistant and graduate student, a year that led to another year in the same capacity. Those two successive years at Nebraska go together, because it was during them that I mounted a series of large skeletons for display in the museum halls.

Phil Orr was away at Chicago and I inherited his mantle. So when Dr. Barbour told me that I was to mount the skeleton of *Diceratherium,* an extinct rhinoceros that once roamed the ancient forested savannahs of western Nebraska, I could only hang on to the edge of a table in his office to steady myself and agree to have a go at it. Such an assignment revealed in Dr. Barbour either a great confidence in my far from established abilities, or perhaps a certain naïveté on his part. I suspect the latter.

Down I went to the paleontological laboratory, to wrestle with *Diceratherium,* the first of four skeletons that I worked on during those two academic years. It was the beginning of a new phase in my education as a prospective paleontologist and museum person.

The setting up of a fossil skeleton is a complex and often difficult operation. In the first place, fossil skeletons are seldom complete, so the missing bones must be modeled in clay and cast in plaster (or nowadays more probably in fiberglass), while incomplete bones must be restored to their complete form. Second, ancient bones are very frequently distorted because of the processes of fossilization, so that allowances must be made for such distortion when the skeleton is assembled. (It is not often possible to remove the distortion; the bones must needs be accepted as they are.) Further, fossil bones are commonly mineralized—permineralized is the more proper term—so that, like glass, they are hard and very brittle. Consequently the mounting of a fossil skeleton entails the construction of a framework of iron or steel to support each and every individual bone in its proper place. This takes some doing, especially if the bones have been distorted to any great degree. In the old days of paleontological techniques the iron supports for skeletons were often cumbersome and conspicuous. Now the idea is to make the supporting structure as inconspicuous as possible.

During that autumn of 1927 I struggled mightily with the skeleton of *Diceratherium.* It was a composite skeleton, made up of bones from more than a single individual animal, and thus there were various discrepancies to be overcome. At an early stage of the game I learned to curse the fact that animals with backbones must necessarily have so many complicated, interlocking bones in their hands and feet, and that they must have so many ribs. The ribs of *Diceratherium* at times drove me almost to distraction. They were slender and delicate, and no sooner would I get one rib put together than it would break apart again, the break usually being at some spot which hitherto had been beautifully whole. Each rib required a strip of iron along its length to support it, and the iron had to be bent to fit the contours of the bone. Of course the rib had to be fastened to the preformed iron. In those days we did not have any of the epoxy resins and other adhesives that are so prevalent now, so fastening the rib to the iron involved the use of wrap-around wires, made as inconspicuous as possible.

Then there was the vertebral column to be strung on a sufficiently strong iron rod, bent to the proper shape. And there were the limbs that had to be supported by irons so bent that they conformed to the intricate curves of the several bones that went together to form each limb. Bending the irons for the limbs was a particularly time-consuming job. As soon as the iron fit in one place, it would be awry in another. And when it was made to fit the second area, it would be out of joint with the first. One had to go over and over the iron, time and time again, fitting and revising and fitting.

Finally the several assembled sections of the skeleton had to be fastened together, to make a reasonable-looking animal. Furthermore, it had to be a skeleton that would stand up and not collapse the first time it received a slight jar.

I persevered, and in time there was a skeleton of *Diceratherium* ready for the exhibition case. It wasn't a very good mount, but it did suffice. There it stood, the visible evidence of a rather small rhinoceros that once inhabited North America, a rhinoceros that unlike modern rhinos had a pair of horns side by side on the nose, not one behind the other.

Then I tackled *Moropus. Moropus* was a chalicothere, an extinct relative of the horses, but an animal with large claws on its feet instead of hooves.

And *Moropus* was a large chalicothere, as big as a very large draft horse.

I put together the four feet of *Moropus*—a pesky job—and then I restored the top of its skull, for the cranium was missing in our specimen. It was at that point that Henry Reider appeared on the scene.

Henry's entrance into my little world was providential, because he furnished the assistance and the skills that I needed in order to go ahead with *Moropus*. Henry, an immigrant from Russia when he was quite small and now thoroughly Americanized, was a sort of midwestern Sam Weller, an unabashed, uninhibited character having many talents. He pitched right in, and soon we had *Moropus* taking form at a rapid and encouraging pace.

One of Henry's accomplishments was ironmongery, and that was invaluable, because *Moropus* was going to require some pretty fancy ironwork to hold it together. We had a forge in the laboratory, a regular old coal-fired forge, with a bellows to blow the flame into high heat. Henry and I spent many hours at the forge, heating iron rods and straps, pounding them into shape on a nearby anvil, and tempering them in a tub of water. Very shortly it was Henry who was doing all of the ironwork, and I was indeed happy to turn over to him that part of the task.

Before we ever started on the iron frame for *Moropus* I had assembled the skeleton on a sort of temporary wooden scaffold, a Rube Goldberg affair that was a marvel to behold. The bones were all posed on this contraption in their proper positions, so that the irons could be fitted to them. Right in the middle of our work one day there was a ringing of bells and a tremendous bustle in the corridor outside the laboratory, and suddenly some large and burly firemen burst into our quarters dragging a hose with them. They headed for a small door on the opposite side of the room that led into a heat tunnel connecting our building with other parts of the campus, and on their way they unceremoniously banged into poor old *Moropus,* which swayed back and forth on its temporary underpinnings like the proverbial ship in a gale. I held my head, expecting to see the skeleton disintegrate in a series of bone-shattering crashes, but *Moropus* held up. The firemen opened the door to the heat tunnel, smoke rolled out, and then I had visions of *Moropus* perishing in flames. Soon, however, the blaze was extinguished (it had developed during some construction work in the tunnel) and the firemen emerged, more quietly this time, to go around *Moropus* in the proper manner.

Moropus was completed without further incident and took his place in the exhibition hall, where he stands today. So ended my second year at the university.

During the next year Henry and I set up two skeletons, one of a giant entelodont, an animal about the size of a bison and related in a very distant way to modern pigs, the other the skeleton of a modern African elephant.

First we started with the entelodont, known as *Dinohyus.* This rare and interesting beast had a perfectly enormous skull, armed with huge canine teeth. Behind the skull was a strong skeleton, with long legs and feet—obviously an animal adapted for running across the land at a pretty good clip. *Dinohyus* was and is one of the treasures of the Nebraska University Museum, and Dr. Barbour was inordinately proud of it.

Consequently, on our first day with *Dinohyus,* when we got it out of the storeroom to take it to the laboratory, Dr. Barbour was in a perfect tizzy. He insisted on lending his hand to the operation, and about the first thing he did was to put the pelvis on a large hand truck, to be wheeled into the laboratory. Then he went off on some other mission, as was a common habit of his. Henry had not been present when Barbour placed the pelvis on the truck, but right after the director left Henry appeared. I was just on the point of putting some sand bags under the pelvis so that it would ride safely to its destination.

"No," said Henry, "leave it just like it is."

We did, and on the way to the laboratory the pelvis neatly cracked into two pieces, as I had foreseen it would. Henry insisted that we leave it

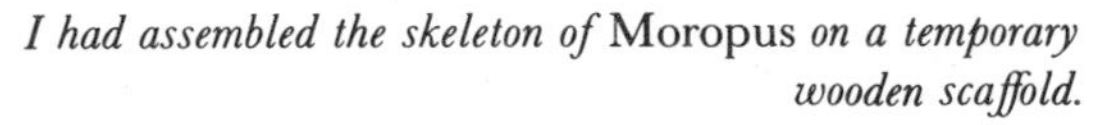

I had assembled the skeleton of Moropus *on a temporary wooden scaffold.*

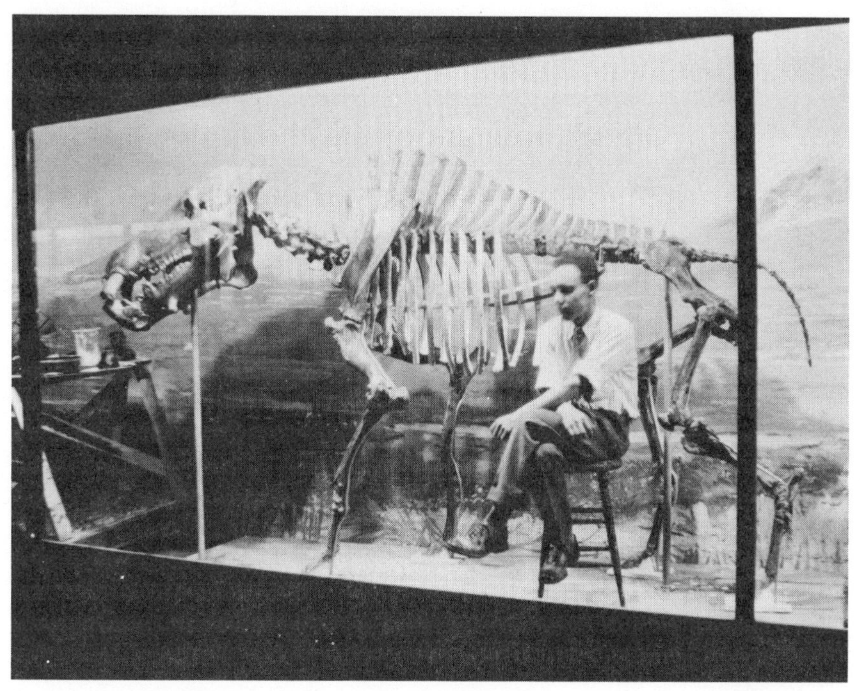

Our mount of Dinohyus *was our best piece of work.*

on the truck when we got it to the laboratory.

At that point Dr. Barbour appeared once again on the scene.

"Look at that!" cried Henry in an indignant voice and pointing to the broken pelvis. "Whoever put that pelvis on the truck didn't have much sense. How could anyone do something like that!" And Henry accompanied his vociferous protest with gestures, making quite an act of the whole thing. Of course he could get away with it, because he could plead ignorance as to who was to blame.

Naturally Dr. Barbour was taken aback. He mumbled some apologies and fled to his office, and from then on we worked on *Dinohyus* in peace. Henry had learned things while setting up *Moropus,* and I had, too. So our mount of *Dinohyus* was our best piece of work to date. It is still to be seen.

Finally there was the African elephant, and being a modern skeleton it was, as the saying goes, duck soup. We were able to put it together in relatively short order.

One incident related to the mounting of the African elephant skeleton may be mentioned. It so happened that I was working on the mount by myself on a Saturday afternoon, in order to push the job along. I was standing on a high stepladder, attempting to bolt an assembled forelimb to the rest of the skeleton. Suddenly the ladder folded, and down I went with a loud crash, clasping the front leg of *Loxodonta africana* firmly against my chest. We landed on the floor with the big leg on top, while down the hall came Dr. Barbour with his coattails streaming out horizontally behind him. As he arrived on the scene and as I was struggling to get out from under the elephant leg, he asked in a concerned voice, "Is anything broken?"

"No," I replied, "I think I'm all right."

"Not you!" he roared. "The elephant!"

Truly there was nothing callous or unfeeling in his query. He was merely showing a museum man's concern with priorities.

From the very beginning of my student career at the University of Nebraska I yearned to go out into the field and collect fossils. Certainly the excitement—one may say the romance—of fossil hunting was one of the attractions of paleontology that strongly appealed to me. Out there on the high plains of western Nebraska were veritable treasure troves of buried fossils, the remains of long-extinct mammals that once had populated what was then a land of lush meadows, forests, and meandering rivers, and I wished with all my heart to go out and find some of those relicts of life long since vanished from the earth. For the summer following my initial year in Lincoln such was not to be the case; the museum just moved into its new quarters must needs get itself organized, and fieldwork for the time being was something a bit too extraneous for immediate consideration. My chance was to come during the next two summers.

Fieldwork is an exciting aspect of paleontology, especially for the young scientist. It is hard and strenuous work, but it has all of the fascination of a search for gold. It is the gathering of the raw data with which the paleontologist is to prosecute his studies, and it is anything but dull. Indeed, a season in the field usually is a time of many adventures, sometimes great, more often small. The collecting of fossils is that part of paleontology especially interesting to the public, so much so in fact that it often obscures the more extended activities of preparing and studying fossils. Although fossil collecting is only the beginning of a long sequence of paleontological disciplines, there can be no denying that it is in many ways the fun part of the science.

In the year before I went to the university two young men in western Nebraska, Morris Skinner and James Quinn, had made a spectacular discovery of fossil mastodonts, the predecessors of but not the actual ancestors of our modern elephants. They also found other large mammals associated with the mastodonts. Skinner and Quinn, being energetic and resourceful lads, set about collecting these fossils, which fortunately were buried on the Quinn farm, not far from Ainsworth, Nebraska. At the same time they got in touch with some museum people, and the upshot was that the fossils, including a beautiful mastodont skeleton, were sold in part to the Denver Museum and in part to the American Museum of Natural History in New York.

It so happened that Childs Frick of the well-known Frick family was intensely interested in the fossil mammals of western North America, and he was associated with the museum in New York. He made arrangements to have Skinner and Quinn collect for him.

All of this made a big impression upon Dr. Barbour, and by the spring of 1928, with the museum safely installed in its new quarters, he turned his thoughts to western Nebraska. He decided that a museum party should go into the field and do what Skinner and Quinn had done, or at least make an attempt in that direction. I was designated to venture into the high plains of north-central and western Nebraska, where in the deeply dissected canyons I would search for fossils. In the first summer of this fieldwork I was an assistant to Professor A. L. Lugn of the Geology Department. The second summer I was in charge of the work, and a young undergraduate, Paul O. McGrew, was my assistant.

(Let us look briefly into the future. Skinner, Quinn, and McGrew all became outstanding paleontologists. Skinner, recently retired, has spent his life with the Frick Laboratories of the American Museum of Natural History in New York. Quinn went to the Field Museum in Chicago and subsequently became professor of paleontology at the University of Arkansas. After his retirement not long ago, he fell from a cliff while looking for fossils in Nebraska with his old colleague, Morris Skinner, and was killed. It was a tragic end to a long and productive life. McGrew, likewise recently retired, spent some years at the Field Museum and then had a long career as a professor at the University of Wyoming. All of which illustrates the fact that, generally speaking, once a paleontologist, always a paleontologist.)

To return to my story, I spent two very busy field seasons in western Nebraska, during the course of which some interesting and important fossils were collected. From the beginning our university museum parties worked in cooperation with Skinner and Quinn, for we saw no reason to be at odds with them. They had many prior rights in the area where they were working, but we were representing the state, so we had some rights, too. The field was very amicably divided between us and there were no problems. Morris Skinner and Jim Quinn were most cordial and helpful to us, and we had some profitable weeks with them, both in 1928 and in the following summer. I should say also that our days with the Skinner and Quinn team were exciting times as well.

Some of the excitement came from the Skinner-Quinn method of transportation. They had a Model T Ford that Morris had purchased for the princely sum of fifteen dollars, and this car was something to behold. It had been stripped of its superstructure, so that it was in effect a chassis with an engine and hood in front, a driver's seat, and behind that, a sort of box to hold groceries and supplies. With that vehicle Morris and Jim would go places that I suspect our modern timid paleontologists would hesitate to attempt with a four-wheel drive jeep.

For example, their mastodont quarry was in the bottom of a narrow canyon, and to reach it they had carved a trail of sorts along the winding creek that sometimes flowed and sometimes did not in the depths of the narrow gulch. But Skinner and Quinn disdained to go to their quarry via the bed of the canyon. Rather, they would drive along the prairie on top, and just before reaching the rim of the canyon would put on speed as they drove head-on toward the edge. Over the brink they would go, the car flying through the air in a magnificent arc, to land on the talus slope partway down the canyon. Then it was a matter of sliding the rest of the way down to the quarry. The first time this happened to anyone riding in the "bug" with Morris or Jim the unsuspecting passenger lost a few heartbeats as well as the loose change in his pockets.

One fine summer day Harold Cook, a well-known paleontologist, came to Nebraska to visit the quarry. He was riding along in the bug with Morris and Jim, talking a blue streak (Harold was a great conversationalist) when he suddenly found himself sailing into the blue like a small plane on take-off. Harold opened his mouth wide but not a word came out. Finally, safely down at the quarry, it took Harold about five minutes to regain his equanimity. It was an experience he talked about with feeling for months, even years, afterward.

Morris used to like to drive the bug by standing on the seat, which gave him a better view of the surrounding landscape, holding the steering wheel with one hand, and waving a .22 rifle with the other, ready to plug any jackrabbit that leaped out of the tall grass. (For the younger generation it should be explained that the old Model T had a hand-controlled lever just under the steering wheel to feed gas; thus the gas lever could be set and the car would rattle along all by itself, as it were.) One day I was riding with Morris through the tall prairie grass (I can't remember whether he was standing on the seat or sitting down in a more conventional manner) when suddenly the front wheels dropped into a hidden ditch. We pitched forward, the tires blew out, and groceries, which were being carried in the box behind, went flying all over the northern Nebraska scene. But that didn't faze Morris; it was all part of his day.

Perhaps my most searing memory of Skinner and Quinn and the bug was the day it got stuck in the middle of Plum Creek, a fast-flowing Nebraska stream some fifty feet or more in width and perhaps three feet deep at this particular place, which was supposed to be a ford. There the bug sat, completely bogged in the sand which was being washed at a rapid rate over the wheels and the underparts of the contraption. Even the genius of Skin-

Imagine my horror at seeing the car flip over.

ner and Quinn could not extricate the vehicle.

Morris hiked out to a nearby ranch and borrowed a team of heavy horses, with which he planned to pull the car out of its sedimentary shackles. The horses were hitched to the front of the car, Morris sat on the hood, and Jim sat at the wheel. Paul McGrew and I stood on the bank, I with a camera to take a picture of the proceedings. Morris shook the reins, the horses put their shoulders to the collars, and I put my finger on the trigger of the camera. Just as I snapped, imagine my horror to see the car flip over, upside down, with Jim pinned on the bed of the stream under two or three feet of sand-laden water. I dropped the camera and dashed for the edge of the stream, but Morris, as quick as a cat, had jumped into the water and was dragging Jim out from beneath the car. Jim emerged waterlogged but otherwise no worse for the experience.

After he had regained his equilibrium, he and Morris swung the horses around, put them to it, and in a jiffy the car was right side up. Then it was dragged back to the bank, where we sat around for a half-hour or so while it dried out. Finally, after Morris had returned the horses to their owner, he and Jim got in their bug and away they went. *Those* were bonediggers, *those* were vehicles, *those* were the days.

Not all of the adventure of being with Skinner and Quinn was experienced in line-of-duty activities; some of it was reserved for the hours of recreation, if so they could be called. One of the favorite Skinner-Quinn games, and I don't know its name, was for Morris to station himself on one side of a little canyon, from which vantage point he would try to keep Jim pinned down by rifle fire, while at the same time Jim, on the opposite side, was doing the same to Morris. On more than one occasion I have crouched behind a rock pinnacle with the one or the other of these two worthies, listening to the bullets whine past, to ricochet off of rocks behind us or to bury themselves in the sandy slopes of the canyon wall. Indeed, I particularly remember one occasion when Morris and I were sheltered behind a big rock, and Morris suddenly thrust his collecting pick out into view with his cap on the end of it. Jim promptly drilled a hole in the headpiece, much to Morris's delight. For an effete easterner from the Missouri Valley like me, it did seem like a strange way to have fun. But that's the way it was, and a good time, more or less, was had by all.

During both of these summers, after having spent some time with Skinner and Quinn, we moved on to the west to work in the general region of Valentine, Nebraska. This was sandhill country, but knifing through the sandhills were steep little canyons that had been cut by the high plains streams as the land was gradually uplifted. It was along the little cliffs and banks of these canyons that we prospected for fossils.

Prospecting was largely a matter of scrambling along the steep escarpments looking for signs of fossil bones. Sometimes one might see a bone protruding from a cliff, sometimes one might find scraps of bones in the talus slope at the bottom of the cliff. Then a decision had to be made. Did the prospect look rich enough to warrant digging into the cliff to try to find more fossils? Digging was arduous work and had to be done carefully. Perhaps a preliminary or exploratory dig would be commenced. Then if it turned up good material a real quarry would be initiated.

Perhaps this is a good place to say something about the methods of getting fossil bones out of the ground. The methods vary somewhat, according to the nature of the fossils, but generally speaking, the process of removing fossil bones from their natural graves where they have rested for so many thousands, or more generally, so many millions of years can be involved and tedious. As has been mentioned, fossils are commonly permineralized, which means that the original hard parts (in this case bone or the enamel of teeth) are infiltrated and frequently recombined or replaced by minerals deposited by ground waters. The fossil, retaining all of the characteristics of the original bone or tooth or whatever other hard part may be involved, even down to microscopic structure, is petrified so that it becomes stone. It is hard, but more often than not it is brittle—as brittle as glass. Therefore one does not dig up fossils like potatoes.

Although that is just what was done in the early days of paleontological fieldwork. Fossils were picked and pried out of the rock in which they had so long been encased, and the remains, usually in hundreds of small fragments, were scooped up into bags, to be carted back to the laboratory for tedious assembly.

As much as a century ago new methods were devised and these have been perfected and augmented through the years. Today very small fossils, such as tiny teeth and bones, are frequently recovered by washing and sieving the broken-down matrix in which they occur, and after that by picking out the fossils from the concentrate obtained, often with the help of lenses or a microscope.

For the larger specimens it is necessary to get them out of the ground intact, and this takes some doing. Initially, after the bone or, more commonly, an association of bones has been located, the specimens are carefully uncovered. This involves removing any rock or soil that is above the fossil deposit—the "overburden"—before the bones themselves can be exposed. When the overburden has been removed to within inches of the bone layer, the fossils commonly are brought to light by the use of hand tools such as hammers and chisels, and little picks or awls, the rock fragments or sand or clay being constantly brushed away from the exposed bones with whisk brooms or soft brushes. As the bones are exposed they are treated with a preservative such as thin white shellac or Glyptol. Then the bones are additionally protected by covering them with thin paper—Japanese rice paper is the best but absorbent toilet paper will suffice—and shellacking the paper to the bones.

When a considerable area of bones is thus exposed, the next task is to map out this fossil floor into "blocks," in preparation for removing each block as a unit. This involves cutting down around the block to isolate it, and this frequently requires the cutting of narrow trenches between the designated blocks if the bones are thickly associated. After each block is isolated as a sort of island on the quarry floor, it is covered with strips of burlap dipped in plaster of Paris, frequently with a separator of damp paper being placed on the block before the plaster bandages are applied. The entire process is analogous to the manner in which a physician puts a plaster cast on a broken arm or leg. Indeed, if time is of the essence, and the fossils are not too large, ready-prepared surgical bandages can be used. Commonly several layers of burlap and plaster bandages are pasted over the top and sides of the block, and if it is of any size, splints may be plastered onto the block to give it strength. These are usually any strong pieces of wood that may be handy—heavy tree limbs or old fenceposts or two-by-fours.

Now the block is ready to be turned over. This involves undercutting it, a particularly difficult task. A large block may be very heavy, so it is undercut in such a way that it rests on a rock pedestal, or perhaps two or three pedestals. Burlap and plaster bandages are often applied to those parts of the bottom of the block that have been undercut. Finally, when the block has been secured by plaster bandages as thoroughly as possible, the pedestal or pedestals can be cut away. Then the block is turned over, with many a prayer that it will not disintegrate in the process. This is the moment when experience and skill count.

When the block is finally turned over, the entire bottom, now uppermost, is plastered, and additional splints are applied if necessary. It is now ready to be shipped to the laboratory, where the entire process is reversed, and the fossil bone or bones can be removed completely from the rock.

By the use of such methods we exposed bones and removed them, for transportation back to the museum. In those days our methods were simple and perhaps primitive. We had no such luxury as Japanese rice paper, so we bought rolls of toilet paper by the gross. So much, in fact, that one day a storekeeper in the little western Nebraska town where we were getting our supplies inquired solicitously as to the nature of our diet when we were out in camp. I suppose, before we had explained the matter to him, he had visions of us suffering acutely from our primitive cooking.

Furthermore, we obtained our burlap in the form of old gunnysacks, which we laboriously cut into strips. Sometimes, we were hard put to it to find enough empty gunnysacks at the various feed stores in town to serve our purpose. But we prevailed, and in one way or another we obtained our supplies and carried on our work.

The work of prospecting and collecting proceeded with adventures great and small. One of my little adventures took place on a hot day when I was prospecting along the Snake River Canyon, south of Valentine. It had been an unproductive morning; not much in the way of fossils had turned up. Along toward noon, I decided to make my way to the rim, a decision that was prompted by the sight of a little path leading up to the canyon's edge. I took the path, and just as I emerged at the top I was confronted by a very large country lady and her numerous children, all armed for battle. The lady flourished a fearsome-looking club, one boy waved a baseball bat, and other children had rocks and sticks in their hands. One more instant and they would have let me have it.

Their defenses went down and they gazed at me with astonishment when my head appeared in full view. They dropped their weapons and with much embarrassment began to apologize. It turned out that having seen me scrambling along the cliffs they thought I was their neighbor, Mr. Collins, sneaking down the canyon to raid their watermelon patch. All was clear, and we sat down and had a good visit.

Later Paul McGrew and I visited the Collins establishment, and we could see why the farmer's wife had her suspicions. The Collinses were real hillbillies transferred to the western plains, and no doubt they would scrounge food in any way they could. They were all ragged, barefooted, and goodnatured. We had an interesting visit with them as well.

That last season we wound up our campaign collecting fossils on the ranch of P. H. Young, a very cooperative landowner. In fact he made available to us a nice little cabin by a windmill, and during our final weeks of collecting we lived in style. Then the season ended; we packed our fossils in boxes to be shipped back to Lincoln and made our way home.

That trip back to Lincoln at the end of the summer field season was the first stage of a longer journey that was to lead me to a new life, very different from the life that had been my lot during the past three years. For during the winter preceding this last foray into the high plains I had made a decision that was as fateful to me as the break in my life three years earlier, when I had taken those first hesitant steps into the realm of paleontology. The decision was to leave Nebraska for a larger university having facilities (not available in Lincoln) for a comprehensive training in paleontology.

During the winter I had applied at three institutions for a fellowship, figuring that I probably would get turned down by all of them, but that I just might get something. As I had feared, my application at the University of California was rejected. Then came a letter from Yale with the same disappointing news. I had about decided that my efforts would lead to nought when I received a letter from Columbia University, advising me that if I would promise not to accept a fellowship or a scholarship from any other institution, Columbia would *consider* me for a fellowship. It was not a definite commitment, but it seemed encouraging, so I sent a letter right off to New York with my promise.

And then on the very next day after I had mailed my letter I received a second letter from Yale saying that my application there had been reconsidered and that I had been awarded a Sterling Fellowship. Here was a pretty pickle! A solid fellowship from one place that I could not accept because of a promise to another place that at best had made only a tentative offer. What was I to do? I made my way to the office of the dean of the graduate school at Nebraska for advice. It was his considered opinion that the letter from Columbia was tantamount to the offer of a fellowship.

He was right. After some days, perhaps a week or two of uncertainty, a second letter came from New York, advising me that I had been awarded a University Fellowship at Columbia. So my course was set.

The little things often make big differences in one's life. If that letter from New Haven had come two days earlier I would have gone to Yale and my whole life would have been quite different from what it was. But there are no regrets. Columbia in those days was perhaps the outstanding place in America to do graduate studies in vertebrate paleontology, especially since one could work under the tutelage of William King Gregory, a scholar of vast erudition whose knowledge of the vertebrates, fossil and recent, was perhaps more comprehensive than that of any other man on the continent. And the University Fellowship was a most generous grant.

Thus I went back to Lincoln from western Nebraska with my mind filled with the wonders of New York, where soon I would be. There was a trip back to Maryville to visit my parents, and then I left for the East with a feeling that ties to my old home and to my former life were becoming increasingly tenuous—that the center of my world was being shifted to a strange and distant locale.

5. LIFE WITH OSBORN

My first year in New York was a time of intellectual discovery and stimulation. The city had much to offer, and I made as much of my opportunities as I could. The theater was my particular delight, and in those days when one could get a good balcony seat for a dollar or less, I spent many an evening in the theater district.

The main focus of my attention was the American Museum of Natural History, where Professor Gregory was located. Years earlier Professor Henry Fairfield Osborn had come to New York to found a department of vertebrate paleontology at the American Museum and simultaneously to establish a program of graduate studies at Columbia University. He instituted an arrangement whereby graduate students at Columbia would do their work at the museum, where they could have access to large collections of fossil and recent animals. It was a good plan for the students, for the museum, and for Columbia, especially because the university would thereby be spared the necessity of trying to maintain large collections for teaching and research. Consequently, I spent much of my time at the museum during that year and the following one as well, although I did put in many hours at Columbia to complete my graduate course work. There I enjoyed the companionship of a closely knit cadre of graduate students.

My horizons were particularly broadened by studies under Dr. Gregory. At Nebraska my experience had been almost entirely with fossils of late Cenozoic mammals found on the high plains: the remains of upland animals that had lived in middle America as recently as a few thousand years ago and as distantly in the past as perhaps 40 million years before the present day. That may seem like a considerable timespan, but it is only a fraction of the history of life on the earth and the fossils of the high plains

The main focus of my attention was the American Museum of Natural History in New York City. (Arrow shows my office.)

represent only a few of the many environments in which animals are evolving through time.

At the American Museum Dr. Gregory spread before us (myself and several other graduate students) the panorama of vertebrate life, from its very beginnings 400 million years or more ago up to the very hour of our classroom lectures and discussions and our laboratory projects. Gregory was an authority of remarkable breadth, so his students were fortunate in that they received instruction embracing the entire range of vertebrate evolution. Dr. Gregory was renowned throughout the world for his knowledge of fossil and recent fishes, he was a recognized authority on the "lower tetrapods," the amphibians and reptiles and especially the important mammal-like reptiles, and his knowledge of the mammals was encyclopedic. He also was famous for his work on primates and on the evolution of man. Moreover, he trained two or three generations of vertebrate paleontologists and anatomists, not the least of whom was Alfred Sherwood Romer, in his middle and later years one of my very close friends.

Assisting Dr. Gregory in his teaching and research was Harry Raven, an explorer and an anatomist par excellence. He had conducted Gregory on worldwide field expeditions to collect animals for anatomical research, and just before I arrived in New York they had gone to Africa to obtain several gorillas, which they embalmed in the field and placed in large tanks, made specifically for this purpose, which were then manhandled through the jungle by an army of porters. Eventually these gorilla cadavers, safely stored at the American Museum, were described in a classic study of gorilla anatomy.

Gregory and Raven made an effective and in some ways a most un-

Dr. William King Gregory spread before us the panorama of vertebrate life. (Painting by Charles Chapman. The portrait is currently hanging in the Columbia University Classroom in the Frick Wing at the American Museum of Natural History.)

likely team. Dr. Gregory was truly rather unworldly; his mind was constantly in the clouds, and I suspect he gave little attention to the small, practical aspects of life. He could not be diverted by such things. Harry Raven was a very practical person; he could go into the far corners of the earth, as he did frequently, and carry on his work efficiently no matter what might be the obstacles and frustrations in his way. One could learn much from those two dedicated men.

In his younger years, around the turn of the century and for a decade afterwards, Dr. Gregory had been a close associate of Professor Osborn, serving as his assistant and collaborator. Then, as Dr. Gregory gained scientific stature, the task of assisting Osborn in his research fell on the shoulders of Charles Craig Mook, a curator at the museum. In time Mook became involved in his own program, so he began to search for another person to take *his* place as Osborn's assistant. Consequently, one day in the spring of 1930 Dr. Mook asked me if I would like to be Osborn's scientific assistant. The academic year was drawing to a close, and I was wondering what I was to do during the coming year, for my fellowship was not renewable. Mook's suggestion came at just the right time, and I accepted. It meant extra work for me, and extra problems, but it did solve the question of my immediate existence.

It was the beginning of a five-year association with Professor Osborn, a time filled with experiences and incidents.

Osborn was then a larger-than-life figure in the field of vertebrate paleontology, as well as in the museum and academic worlds. He was the author of a prodigious number of paleontological contributions (made possible in no small degree by the coooperative labors of his colleagues and assistants), concerned especially with fossil mammals and to a lesser extent with fossil reptiles. He was particularly interested in the evolution of man, and was the center of many controversial debates on this ever-changing and sensitive subject. He was president of the American Museum, and a professor emeritus at Columbia. He was a recent past-president of the American Association for the Advancement of Science, and the honors and degrees he had received were legion.

Professor Osborn was quite aware of his eminence. I had seen him at a distance in the halls and corridors of the museum—a large and forceful-looking person, generally clad in rather formal clothes, of which a vest with a white piping along its edges was a noticeable feature. He would often be seen striding along with two or three people in his wake, like a majestic ocean liner accompanied by several gulls. It was a formidable sight, and I wondered what I was in for. I was soon to find out.

I should mention by way of anticipation that in addition to being an eminent man of science, Osborn enjoyed the advantages of considerable inherited wealth. He had a home at Garrison-on-Hudson, high on a hill

Dr. Henry Fairfield Osborn was a larger-than-life figure in the field of vertebrate paleontology.

across the river from West Point, a sort of New World castle on the Rhine. (I think he also had a townhouse in New York.) Castle Rock, as his place up the Hudson River was called, was his post of command from which he sallied forth daily for his train trip to New York.

It was never my fortune or misfortune to have to go to Castle Rock, but the Professor's devoted scientific secretary, Mabel Percy, and his editorial secretary, Ruth Tyler, would make frequent trips there, especially in the summer. I got the impression that those were strenuous journeys—up at dawn, to Grand Central Station in Manhattan, usually lugging a heavy bundle of manuscript and proof, arrival at Garrison, taxi to Castle Rock, and work all morning in a little sort of gazebo or summer house down the path from the castle. Then back to the city in the late afternoon, arriving hot and tired, while the Professor relaxed on the cool heights above the majestic Hudson River.

From this it will appear that Osborn was a bit imperious, and he was. But one had to take him in context. He had grown up in the lush days of the Robber Barons, and he viewed society as a highly stratified arrangement, in which he occupied a top stratum. Yet he had a genial side and, according to his own lights, a real concern for people. I know that he was interested in my scientific progress and was very willing to further it—as long as I performed my duties as his scientific assistant.

The performance of such duties had certain disadvantages. I felt that as long as I was working with Osborn, I should tell him things as I saw them—and I did. That did not always please him, because he did not like to have his firmly held concepts disputed. Consequently, we had some pointed, if not heated, arguments that would have been very entertaining to any bystander. I know he felt that I was very impertinent to oppose him, yet I think he respected me for it, just the same. Life with Osborn certainly could be trying at times, but it was interesting.

Not long after Dr. Mook had spoken to me about signing on as Osborn's assistant he introduced me to the legendary figure. Osborn greeted me cordially and we had a pleasant conversation. He told me how he used to study the works of Edward Drinker Cope (he was a disciple of Cope and was personally acquainted with the famous Quaker naturalist) and he talked about how virtuous it was for young men to study the works of their elders. He made it clear that I would greatly profit by studying the works of Osborn. And that is how it began.

Osborn liked to work on big things, and he was then doing a monographic study of fossil elephants. One of my first tasks was to go down into one of the exhibition halls at the museum to help him measure the skeleton of a large mammoth. Rachel Husband Nichols, a member of the paleontology department, was dragooned into helping on the job.

We installed a card table beside the skeleton and Professor Osborn sat himself down. With the aid of a ladder I swarmed all over that skeleton, measuring various bones and combinations of bones with a set of giant calipers. I would measure a bone, call down the measurement to Rachel, who would repeat it to Osborn, who, as he wrote down the figures, would repeat it again in a stentorian tone that reverberated through the hall, the sound bouncing back and forth between the exhibition cases and through the free mounts of skeletons in the middle of the room. I think he enjoyed the sound.

Professor Osborn and I visited that hall on numerous occasions, because that was where the elephants were. One day as the Professor (also known to us as HFO or Uncle Hank) and I were walking through the hall we met a tall, genial man with a shining bald head. He was leading a small child, obviously his grandchild, by the hand. When he and Osborn spied each other they exchanged cordial greetings and Osborn introduced me to the visitor. He was none other than Charles Dana Gibson, the famous illustrator, creator of that belle of the turn of the century, the "Gibson Girl." We had a nice little talk and then went on our way.

On our way took us to the elevator, where Osborn, as was his custom, called out the commanding word—"Up"—and waited for the lift to arrive. That was Osborn's system, and the elevator operators were well acquainted with it. Osborn would watch the pointer above each elevator door which showed the progress of the car in that shaft, and then he would sing out before the nearest car reached his floor, whether it was on its way up or down. And if the car was going in the way opposite to Osborn's desired direction, the astonished museum visitors in the elevator would be given an unexpected ride. A simple system for one in authority, and having the commanding voice of authority.

Professor Osborn had at this time become much intrigued by the increase in number and height of the enamel plates in the cheek teeth of fossil and recent elephants. It should be explained that the molar teeth in elephants are massive affairs, consisting of tall plates of enamel, each plate rising from the base of the crown to its grinding surface where, turning the corner, it descends again to the base of the tooth. One might describe each plate as a broad ribbon of enamel going from base to occlusal surface and back to base, enclosing a wide pillar of dentine within the narrow space between the ascending and descending enamel bands. In a large elephant tooth there may be two dozen or even thirty such plates, one behind the other, all of them buried with a heavy coating of tooth cement, which is the outer bony tissue covering the tooth in various mammals. If one measures the cumulative length of the enamel in a longitudinal section of one large molar the results can indeed be impressive. In fact, among some of the late Ice Age mammoths the total enamel length in a third molar may exceed thirty feet.

But in some of the early Ice Age mammoths the teeth are not so high and there are fewer plates, so the cumulative enamel length is much shorter. Osborn came up with the idea that one could measure the progress of late geologic time by measuring the length of the enamel in fossil elephant molars. Short enamel lengths were characteristic of early Pleistocene or Ice Age mammoths, and long enamel lengths of late Pleistocene mammoths.

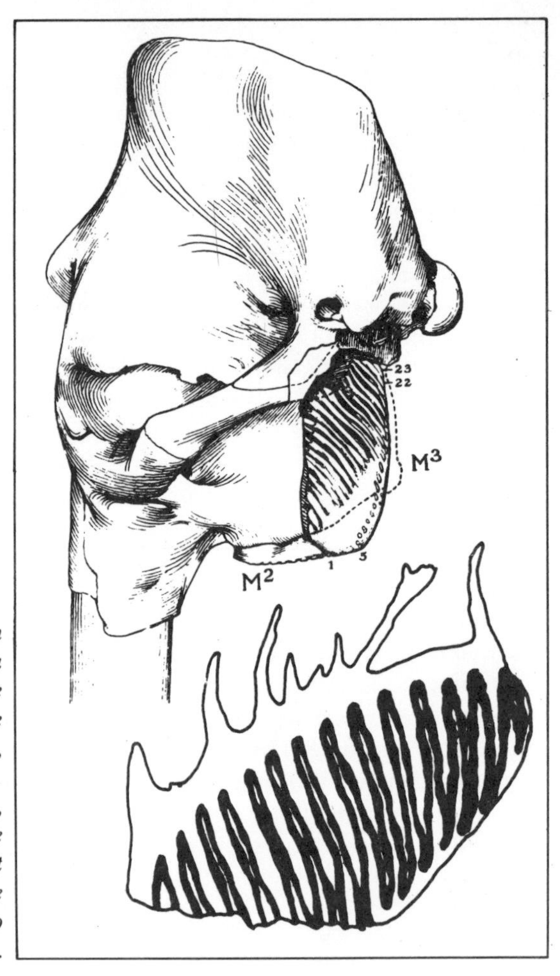

Below is a section of an upper third molar, M3, of a fossil elephant, showing the enamel as black lines. The wear surface, at the bottom, cut across the enamel plates. Above is an elephant skull, showing the positions of the second molar, M2 (almost worn away), and of the third molar (coming into position).

Therefore Osborn concluded that by extrapolation intermediate enamel lengths indicated mammoths at various stages within the Pleistocene epoch, according to the relative length of the enamel. I did not entirely agree with Osborn's thesis, but it was up to me to measure the teeth. Consequently, I spent many hours with a little wheeled map measure, going up and down the enamel bands of sectioned elephant teeth, and adding up the results.

The Professor was impressed; so much so that he decided to have the lengths of the enamel bands in third molars of various fossil elephants plotted full size as long charts. (He could very well have had the job done to scale as small graphs, which could have been made into lantern slides, but that was not the Osbornian way.) With much effort and the cooperation of some museum illustrators the charts were produced.

With such impressive charts at hand, Osborn decided that there had to be a talk and a demonstration somewhere, and what could have been a more suitable forum than the distinguished American Philosophical Society in Philadelphia. Arrangements were made for the Professor to appear at the spring meeting of the society and for me to do the honors as his assistant. On a fine day in April of 1931 I went to Philadelphia on the train, lugging the charts with me, while Osborn came down from New York in his car, driven by his chauffeur. We met at the Philosophical Society's building, a fine old colonial structure near Independence Hall, with some time to spare.

One of the first people we ran into was Professor William Sinclair of Princeton. Sinclair was always an outspoken person; he had some knowledge of what Osborn had been doing with elephant molar enamel, and he didn't have a very high opinion of the business. He was not one to mince words, so he started right in on HFO, giving the Professor a bit of his mind in no uncertain terms. But Osborn was feeling very amiable that day; he let Sinclair's remarks go in one ear and out the other and smiled sweetly all the while.

When the time came for Professor Osborn to give his paper he got up and told the assembled group all about elephant enamel and geologic time, while Sinclair sat over on one side muttering to himself. Finally Osborn reached the point where he wanted to show the charts.

"Mr. Colbert," said HFO, "would you please take one end of this chart, and Professor Sinclair, would you take the other end? That's splendid!"

There was Sinclair across the room, holding one end of the chart, with me at the other end, and Osborn in the middle, talking away as happy as could be. All the time the Professor was talking I could hear Sinclair at the other end of the line grumbling along in a sort of obbligato, a background accompaniment, one might say, to the main theme of the fugue.

After the meeting was over Osborn and I came out of the building and he hailed his car. We climbed in and he instructed the chauffeur to take us to the Academy of Natural Sciences. Off we went, and about the first thing HFO did was to tell his chauffeur to go the wrong way down a one-way street. The driver expostulated but the Professor insisted, so down the street we went. We soon encountered a copper, who in the tactful manner so characteristic of metropolitan policemen wished to know what we thought we were doing, et cetera. HFO stuck his head out the car window and tried to reason with the minion of the law, even identifying himself. Alas, it was to no avail! The cop made us turn around and go back the way we had come, assuring us all the while that he had half a mind to take more drastic steps.

That took some of the edge off our journey to the Academy, and during the rest of the drive HFO gave me quite a lecture about what uncouth people most policemen are. When we arrived at the Academy, Professor Osborn, still out of sorts, was additionally disturbed to find the museum deserted. It was a Saturday afternoon, a fit time for the custodians and the curious public, but not for the staff. The Professor was nettled because he had planned to present an autographed copy of his book on Cope, just off the press, to the Academy brass. The ceremony accordingly was reduced to a prosaic matter of giving the book to the man at the door, with instructions about passing it on to the proper person.

By this time the Professor was ready to go home, so he told me how to get to the railroad station (which I could see, down the Parkway) and then off he went with his chauffeur through the murk of the late afternoon.

With this occasion behind him, HFO decided that elephant enamel should have a place at the meetings of the British Association for the Advancement of Science, scheduled for the late summer of 1931. He and I would go to England in August, measurements of elephant enamel would be made at the British Museum, the results would be added to those already obtained, and then the Professor would expound on all of this at the BAAS (British Ass to the disrespectful). It was the centenary meeting of the association, and big things were planned.

Dr. Gregory also was going, as a companion to Professor Osborn as well as in his own right. Uncle Hank had it all worked out; he and Gregory would travel first class on one of the big liners and I would travel third class, and every day HFO would invite me up into the rare environs of the posh travelers to share with him thoughts about elephant enamel. I didn't think much of this plan, and I doubt that the steamship line would have looked upon it with favor. At any rate, I argued my way out of it (I don't know just how) and arranged to sail to England on a modest cabin-class steamer. Osborn and Gregory left in splendor and I took off separately—being given a rollicking farewell at the ship by a group of museum colleagues, including Margaret Matthew, who had come to the museum at the beginning of the summer as a paleontological illustrator.

How can I write about the magic of that first trip abroad? It was a new and wonderful experience, never to be forgotten, and it was a very special experience, in part because of the circumstances that went along with it. People did not travel so casually in those days as they do today, when airplanes take off every few minutes from numerous airports for journeys across oceans and to distant lands. A steamship departure was quite an event, and a trip across the ocean, lasting a week or more, provided an interlude, a separation in time and space, that made the arrival at a foreign port a truly exciting occurrence. As I arrived in England I had the exciting impression of being in a land very different from what I had previously known; it was new and strange, much more so than is the case today when countries and cultures are homogenized by the frequency of commonplace air travel and the ubiquity of modern communication.

In London, Osborn and Gregory settled themselves in Mayfair, while I found digs in Redcliffe Gardens, just off Earl's Court and convenient to the South Kensington or natural history division of the British Museum. Having journeyed to London by our separate ways, HFO and I met at the South Kensington Museum, an impressive Victorian structure of immense length and unpredictable architecture. It was a damp, rainy day, and even though the month was August the air was chill. The authorities at the museum made Professor Osborn comfortable in the semi-basement room where he had been installed by providing an electric heater, which was placed at his feet. They also gathered together the many mammoth teeth in the British Museum collections, quite a few of which had been longitudinally sectioned, and laid them out on tables in the room. HFO and I got down to business, examining the teeth and discussing their characteristic features.

This went on for a day or two, but then the Professor had other things to do, so he left me to wrestle with the details. I measured enamel lengths, and in the course of time I had quite a body of data supplementary to the figures we had obtained at the American Museum. Therefore Osborn decided to make up some new and more comprehensive charts—again at full scale—which meant great rolls of paper thirty feet and more in length.

With this in mind he said to me one day, "Mr. Colbert, I want you to go over to the Imperial Institute and ask the director to assign one or two draftsmen to you, to make these new charts." Or words to this effect, probably expressed in tones more commanding than I have indicated. I tried to tell the Professor that I could not do this alone, that the request needed to be put forward by someone with his authority.

"Nonsense!" he said. "Of course you can do it. Just go over and tell them that I want the charts."

Again I demurred, and by dint of some persuasion I finally convinced him that the request should come from him in person. We made arrangements to go to the Imperial Institute the next day, to see the director.

The next day we were ushered into the office of the director. He was most cordial and why should he not be to a man of Osborn's position? The director promised all possible cooperation and told me to come over the following day with my figures and we would get to work.

As Osborn and I went down the steps of the institute to the street he turned to me with a smile and said, "See how easy it was!"

The following day I went to the institute and began work with the draftsmen, feeling like a fool as I asked these men, much older than I was, to stretch great long rolls of paper on the floor and get down on their hands and knees to do the necessary drafting and lettering. But they accepted the task with good humor, and eventually it was completed, finished in India ink and watercolors.

Later, at the session of the British Association where Osborn gave his talk, we pinned the thirty-foot-long chart high on the wall where all could see it; there was no problem this time of the chart being held up by willing or unwilling hands. Osborn expounded at length as to the significance of the chart, but I am not sure that his remarks were completely comprehended. For at the end of the lecture some old boy got up and said he understood that Osborn had studied under Thomas Henry Huxley in these very precincts. This pleased HFO ever so much. He *had* studied under Huxley and was very proud of the fact. Moreover, he liked to tell about the day when Darwin, on a rare visit to London, came into Huxley's laboratory, where Osborn and other young men talked with the great man. Of course on this day he repeated the tale, and for the duration of the sessions it was Old Home Week with reminiscences of Huxley, while elephants were completely forgotten.

Elephant teeth were not the only reason for my being in London. Some months previously arrangements had been made at the American Museum for me to study a large collection of fossil mammals from the Siwalik beds of India, made in the early twenties by Barnum Brown of the museum staff, for my doctoral thesis, then in its formative stage. (The thesis eventually was completed, and I was awarded my degree in 1935.) The British Museum had another large collection of Indian fossil mammals, so we had agreed that after Professor Osborn returned home I would stay on to make comparative studies of these fossils. And I did. Furthermore, I crossed the channel in October (on an old Vickers trimotor plane) to visit various museums and collections in France, Belgium, Germany, and Switzerland. For me it was quite an autumn and winter.

In London I became well acquainted with W. E. (Bill) Swinton, at that time keeper (equivalent to an American curator) of fossil reptiles at the British Museum. We had some grand times together and developed a friendship that has lasted through the years. Bill is a person of exceptional charm, his mind is filled with delightful conceits, and the hours and days spent with him have been memorable in many ways.

Day after day at the museum I would study Siwalik mammals—deer and antelopes, pigs and hippos, elephants and giraffes, and lesser animals that once inhabited the plains and forests of ancient India. In the evenings I would perhaps have dinner with Swinton, or go to the theater, or go to my room in Redcliffe Gardens, where I would stick a shilling in the gas heater and try to keep warm.

In the course of things I was invited to the home of Sir Arthur Smith Woodward at Haywards Heath in Sussex, to the south of London. I was

In London I became well acquainted with W. E. (Bill) Swinton.

there twice: once in October, when the rolling hills were clad in the russet colors of autumn, and once at Christmastime, when I spent a week with the Woodwards. Sir Arthur was at that time the grand old man of British paleontology, having retired a few years previously from the British Museum. It was a privilege for me, a very raw recruit in the paleontological field, to be able to spend days at Hill Place, the Woodward home, where I could talk with Sir Arthur. It was all very pleasant, and in some ways faintly Wodehousean. Of course I did not attempt to pinch any silver cow creamers in the middle of the night, but I did have frequent run-ins with the Woodward dog, Binky, who never seemed to appreciate the fact that I was a guest in the house.

The fall and winter rolled by, and in January the time approached for me to take the boat back to New York. Unbeknownst to me Professor Osborn had made plans during the late fall to go around the world on a cruise ship, the *Stella Polaris.* He wanted me back before his departure, and with characteristic Osbornian assumption he cabled to me in London in care of Thomas Cook. But Thomas Cook did not have me on their list, and I never received the cable. (Why he didn't cable me at the British Museum, I shall never know.) Consequently, I proceeded in blissful ignorance of Osborn's plans, with the result that he left on his trip before I arrived back in New York.

When I did get to the museum I found a feeling of merriment pervading the place because of a newspaper story that had come out just before the Professor departed on his trip. It concerned the ship on which he was to sail, and the gist of the account had to do with the immense amount of alcohol on the ship when it docked in New York. (Remember, this was during the days of Prohibition.) The headline read "Cruise Liquor Stock Staggers Customs Men." Then the story went on to tell how the customs officials, who went on board to inspect the ship and who were used to seeing large amounts of liquor, were "a bit staggered by the 'tween deck stores." "The customs men estimated that each passenger may drink about 275 bottles of wine, twenty-five quarts of spirits and as much beer as he can manage in the 100 days" (of the cruise). Finally, "Among the passengers . . . are Professor Henry Fairfield Osborn, president of the American Museum of Natural History. . . ."

That winter passed into spring while I continued with studies on fossil elephants according to written instructions that Osborn had left behind, instructions which were amply sufficient for the purpose, and alternatively worked away on my studies of Siwalik mammals. I wondered sometimes how the Professor was making out on the cruise amidst the incredible stocks of alcoholic beverages and boozy passengers. Evidently he was making out all right, because he returned at the end of April much refreshed and replete with tales of the places he had been and the things he had seen.

Professor Osborn must have had some interesting tales of his young years, but he never passed them on to me. Perhaps he felt that our time together was too restricted for storytelling; we must keep after the elephants. When he was a young Princeton undergraduate he journeyed, with two fellow students, William Berryman Scott and Francis Spier, to the Bridger Basin in Wyoming to search for fossils. That was in 1877, just after the Battle of the Little Bighorn, and in the days when there were some of the old-time mountain men still around. A picture taken at the time shows Osborn, Scott, and Spier posed in their field clothes, Scott holding a large and efficient-looking revolver in his hand and Osborn with a heavy carbine across his knees. I have often wondered if Osborn ever pulled the trigger.

Professor Scott, who spent his life at Princeton, once told me a story of the 1877 expedition. One day the young paleontologists met one of the old mountain men, after they had been out in the badlands. Osborn had managed to sunburn his large and prominent nose to a bright red hue. The old mountain man looked at him and said, "Young feller, either you've got to stretch out the brim of your hat or pull in your nose!"

In the summer of 1932 Dr. Guy E. Pilgrim came from England to study some of the Siwalik mammals in the American Museum collection. Pilgrim, recently retired, had been for many years a member of the Geological Survey of India and had an intimate knowledge of fossil mammals of that subcontinent. He was a pleasant person, if a bit foggy at times. He arrived just in time for one of the monthly meetings of the Osborn Research Club. That was an institution of the museum scientific staff. The group would meet monthly in the museum's Osborn Library (a branch library devoted to vertebrate paleontology), where some staff member would present a paper or give a report on his or her current research. Professor Osborn would sit in a large armchair in the middle of the front row.

On this occasion HFO, always one to indulge in the grand gesture, got up and gave a little speech of welcome to Dr. Pilgrim. At the end of his talk he held out his hand and said, "Dr. Pilgrim, we take great pleasure in welcoming you to the American Museum of Natural History."

Pilgrim had been sitting quietly in the front row puffing at the stub end of a cigar. He got up and went forward to shake hands with Osborn, with the cigar stub held between the second and third fingers of his right hand. Osborn firmly grasped the Pilgrim hand and of course got the hot end of the cigar in his palm. There was an immediate cry of surprise and pain, HFO jerked his hand back, as did Pilgrim, and the cigar, describing a glowing, red arc through the air, landed on the floor. No one could contain himself and a shout of laughter went up.

The Professor carried it off well. "You see," he said, "Dr. Pilgrim is burning with enthusiasm!"

Still my work continued, on Indian fossil mammals and on elephants. All of this time Margaret Matthew and I had become increasingly friendly, in time we fell deeply in love, and finally, early in 1933, we decided on marriage. Margaret is the daughter of W. D. Matthew, a noted paleontologist who spent much of his life at the American Museum as a colleague of Osborn. In 1927 Matthew resigned his museum curatorship to go to Berkeley, where he had been offered the chairmanship of the newly created Department of Paleontology at the University of California. His tenure at Berkeley was short; in the summer of 1930 he died of a kidney ailment. Margaret, who had been trained at the California College of Arts and Crafts in Oakland, came to New York in 1931 as a scientific illustrator.

Osborn was much interested in our forthcoming marriage (which took

Our marriage took place on July 8, 1933.

place on July 8, 1933) and he gave us a tea set as a wedding present. It had about the fanciest teapot imaginable, one that could be set in one position for steeping the tea and then could be tipped back to raise the tea leaves out of the water. With the tea set there was a handwritten Osbornian letter, instructing us how to use the teapot and full of good advice on the advantages of drinking tea rather than coffee. That was just one of the accumulation of Osbornian notes I received, dealing with life, fossils, and the decline of dedicated people in this modern age.

The years passed. Then on a November morning in 1935 Charlie Lang of our preparation staff came down to the fourth floor, where I was trying to plan a little exhibit, to inform me that HFO had died quietly only a couple of hours previously. He had gotten up and dressed and had died after breakfast while sitting at his desk in his home in Garrison.

My life with Osborn, the man, had ended. But my association continued, for I carried on with fossil elephants, to help complete the massive monograph on the Proboscidea on which Osborn had been engaged for so many years. But from then on I could concentrate more and more on my own research interests, which in those days were centered largely upon the fossil mammals of Asia.

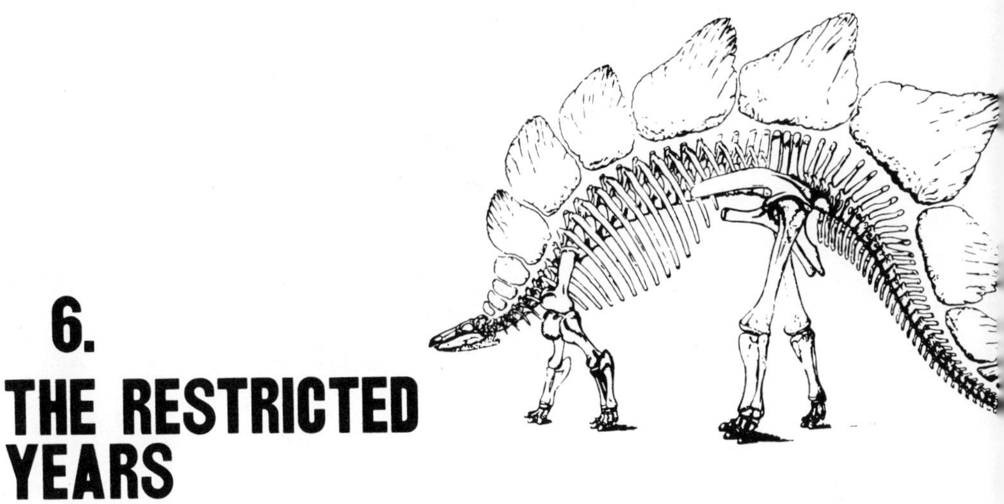

6. THE RESTRICTED YEARS

My interest in the fossil mammals of Asia was in the beginning largely fortuitous. The Siwalik mammals of India were an introduction, and a very large one. Work on Indian mammals stimulated an interest in ancient mammals from other parts of the great Asiatic continent.

It so happened that the American Museum had large collections of fossil mammals from Mongolia, brought together by the well-publicized Central Asiatic Expeditions, which had worked in the Gobi during the twenties under the leadership of Roy Chapman Andrews. There were five such expeditions—in 1922, 1923, 1925, 1928, and 1930. Then the work came to an end because of political unrest in China. As a result of these expeditions there was a treasure trove of Mongolian fossil mammals in the drawers and on the shelves of the museum storerooms, waiting to be studied.

Dr. Matthew, who would have studied many of the fossils, had left the museum in 1927 for the University of California. His replacement, George Gaylord Simpson, a brilliant paleontologist who was to rise with meteoric speed to become during the next half century a world authority on fossil mammals and a profound student of evolution, was largely interested in the very ancient mammals of North America. The later mammals of this continent were at that time being collected and studied by Childs Frick, a museum trustee who had established a special laboratory at the American Museum for this purpose. I felt a squeeze so far as North American fossil mammals were concerned, and so I turned to what was available—the fossils collected in Mongolia. It was not my fortune to do any fieldwork in Mongolia; such work ended just as I came to the museum. But in Walter Granger, Curator of Fossil Mammals, who was the paleon-

Walter Granger and I are watching the antics of my oldest son, George, and a playmate. When I knew Granger he was in middle age, a large, very friendly man with a hearty laugh.

tologist on the Mongolian expeditions, I had a wise and sympathetic advisor. He filled me in on details of localities and with stories of field experiences. In this way I acquired some background for my studies on the fossils.

Walter Granger was a wonderful person. He had come from Vermont to the museum as a very young man, and the museum was his home for all of his adult life. For many years he was a colleague and a close personal friend of Matthew, and they collaborated on many scientific projects. When I knew him he was in middle age, a large, very friendly man with a hearty laugh. Very quickly he became a sort of Manhattan father to me, a friend to whom I could go with my problems, scientific and personal. He loomed large in my life for some eleven years, until the day of his death in 1941.

Those were the Depression years, when life was austere for many people and promotions were limited. In 1933 I had been appointed as Assistant Curator in the Department of Vertebrate Paleontology at the museum, a post I was to hold for nine long years, until I was advanced to the status of a full curator in 1942. (I remained a curator at the museum until my retirement, serving for some years as Chairman of the Department of Fossil and

Recent Reptiles and Amphibians, and subsequently as Chairman of the Department of Vertebrate Paleontology.)

The museum could carry along its scientific program, but on a very modest scale. In those days there were no such things as government grants to aid research; any activities beyond the usual ones depended upon help from interested citizens who had money to contribute. Alas, I had no wealthy friends to support the more esoteric aspects of my research, so I spent my time within the museum walls, studying fossils and preparing papers for publication. That kind of work could be carried on within the regular museum budget. Fieldwork, the gathering of raw data on which the paleontologist depends, seemed to be out of the question, so I continued with Asiatic fossils.

As the months passed by and became years Margaret and I did feel ever increasingly the strictures of trying to get by on my small salary. Then, quite providentially, relief came from Philadelphia. The Academy of Natural Sciences there had plans to revitalize its program in paleontology. There was no money for a full-time curator of vertebrate paleontology but something might be accomplished on a part-time basis. The director of the academy, Charles Cadwalader, approached me to find out if I would be available to help. Walter Granger was most cooperative; he felt that I could have a day off each week from the museum with no loss of pay, in order to go to Philadelphia for consultation at the Academy. It was, in effect, a way for the museum to give me a small raise with no strain on its budget.

Thus began my years of once-a-week commuting to Philadelphia, lasting from 1936 until the opening months of the war. Every Monday morning I would rise very early, get to Pennsylvania Station for the eight o'clock train, arrive in Philadelphia at nine-forty, spend the day at the Academy, and then get the five o'clock train back to New York, arriving home for a late supper. It was a strenuous program, but an interesting one. In Philadelphia I worked with several people, especially Dr. Edgar Howard, who carried on many of the details of the program that we jointly planned; Dr. Benjamin Franklin Howell of Princeton, who was advising the academy on its program dealing with fossil invertebrates, just as I was with the vertebrates; and Dr. Horace Richards, paleontologist at the academy. We had some interesting times together. Moreover, for a couple of years I conducted a course in paleontology at Bryn Mawr College.

All of this was more than helpful to our domestic arrangements. Our oldest son was born in 1937, and a few months later we moved from Manhattan to Leonia, New Jersey, just across the George Washington Bridge. Leonia was to be our home for thirty-two years, and there our five sons grew to maturity.

Since there was no chance for me to go to Asia to carry on fieldwork supplementary to my Asiatic studies I had to depend, as has been said, on advice and help from Walter Granger, as well as other people who had been there. One of the other people was Père Teilhard de Chardin, the famous French paleontologist and philosopher. Teilhard had made a collection of fossils in Burma, in company with Helmut de Terra, a geologist of renown, and the collection was turned over to me for study. As a result I had a close acquaintance with Teilhard during the times when he was at the museum in New York. He was an interesting and a charming person, and it was a pleasure to talk with him—to discuss paleontological problems together. Little did I realize in those days that Teilhard, whom I regarded as a paleontologist with much experience in Asia, would someday become revered throughout the world for his philosophical and ethical writings. (I must admit that I have never really understood this aspect of his work, but I must also admit that many people are impressed and inspired by the work of Teilhard. So much so, in fact, that a sort of Teilhard cult has grown up.)

At last I had an opportunity to get back briefly into fieldwork, from which I had been deprived for nine years. The year was 1938, and the Academy of Natural Sciences in Philadelphia decided that something more or less spectacular was needed for the new paleontology hall being developed there. We talked it over and decided that a large block of Lower Miocene mammals from Agate, Nebraska, would be most appropriate.

Agate was and is famous as a paleontological site—so famous, indeed, that it is now a National Monument. But in 1938 the Agate site, a couple of buttes known as Carnegie Hill and University Hill, rising from the prairie, was still very much a part of the Cook Ranch. Which leads to a tale.

James H. Cook—Captain Cook to all who knew him—was born in southern Michigan in 1857. There he grew up, and there as a lad he saw large flocks of passenger pigeons—now long since extinct. As a young man he went to Texas to begin a long and exciting life in western North America. He was a cowboy, an Indian fighter and at the same time a friend of the Indians, a hunter and trapper, an army scout, a rancher, and a reader of books. His book *Fifty Years on the Old Frontier* is a most fascinating account of the Old West as it really was, as distinct from the Hollywood or TV versions.

In 1886, following his marriage, Captain Cook purchased from his father-in-law a ranch in the Niobrara Valley of northwestern Nebraska. This became the Agate Springs Ranch, and here Captain Cook built his home, a spacious house in the valley, near the headwaters of the Niobrara River. It was a treeless land, so Captain Cook during the first years in his new home planted a veritable grove of trees, keeping them alive when they were small by going around each evening on a wagon loaded with barrels of water and carefully giving each tree the necessary drink to ensure its survival. In time the trees grew large, so that in 1938 the Agate Springs Ranch was a beautiful high plains oasis—a cool, wonderfully peaceful spot in which to relax

Agate was and is famous as a paleontological site (University Hill on the left, Carnegie Hill on the right).

and talk and look out beyond the trees to the hot, burning plains. It was there that I heard many stories from Captain Cook, then an old man. And it was hard to believe that this elderly, truly refined gentleman had lived the adventurous and exciting life of which he told us. It was a life of hunting and ranging for cattle, of gunfights, of Indian wars (he participated in the Geronimo campaign), and of hardships almost beyond belief. Captain Cook was a true friend of the Plains Indians; he was fluent in the Sioux tongue and was a close personal friend of Red Cloud, the famous Sioux chieftain, and of such well-known Indians as American Horse, Little Wound, and Young-Man-Afraid-of-His-Horses. For many years in his later life Red Cloud and his Sioux family, with a band of Sioux followers, would come and camp in the grove at the Agate Springs Ranch, for long conversations with Captain Cook.

Just before his marriage, Captain Cook was visiting his fiancée at Agate Springs (then owned by her father) and they would go horseback riding across the surrounding country. It was one one of these rides that the Agate Springs fossil quarry was discovered. Here is the story in Cook's words.

"Riding one day along the picturesque buttes which skirt the beautiful valley of the Niobrara, we came to two high conical hills about three miles from the ranch house. From the tops of these hills there was an unobstructed view of the country for miles up and down the valley. Dismounting and leaving the reins of our bridles trailing on the ground—which meant to our well-trained ponies that they were to remain near the place where we had left them—we climbed the steep side of one of the hills. About halfway to the summit we noticed many fragments of bones scattered about on the ground. I at once concluded that at some period, perhaps a year back, an Indian brave had been laid to his last long rest under one of the shelving rocks near the summit of the hill, and that, as was the custom among some tribes of Indians at one time, a number of his ponies had been killed near his body. Happening to notice a peculiar glitter on one of the bone fragments, I picked it up, and I then discovered that it was a beautifully preserved piece of the shaft of some creature's leg bone. The marrow cavity was filled with tiny calcite crystals, enough of which were exposed to cause the glitter which had attracted my attention. Upon our return to the ranch we carried with us what was doubtless the first fossil material ever secured from what are now known to men of science as the Agate Springs Fossil Quarries."*

Captain Cook, being an enlightened man, made his discovery known to paleontologists, with the result that through quite a number of years various institutions—the American Museum of Natural History, the Carnegie Museum of Pittsburgh, Yale University, Princeton University, the University of Nebraska, and Amherst College, among others—carried on excavations at the Agate Springs Quarries. Captain Cook and his family were always generous and delightful hosts to the visiting paleontologists.

It was almost inevitable that Captain Cook's son, Harold, should become a paleontologist, and in particular an authority on middle and later Cenozoic mammals. It was my privilege to know not only Captain Cook, but also Harold and his wife, Margaret, and Harold's four daughters.

With that bit of history behind us, I return now to the summer of 1938, and my preparations to collect fossils at the Agate Springs Quarry. Of course it was not a field program intimately related to my research on Asiatic mammals, but it was a part of the Academy effort in which I was strongly involved. It promised to give me a change of pace and some time in the open air, and I did look forward to that. The negative side of the plan was that I would have to desert Margaret and our one-year-old George for a few weeks, but she was cheerfully willing to put up with that.

* James H. Cook, *Fifty Years on the Old Frontier* (Norman: University of Oklahoma Press, 1957; reprint of original edition, published in 1923 by Yale University Press), p. 234.

The daughter of a paleontologist, she thoroughly understood the demands of the profession.

We were to be a party of five: Edgar Howard; an elderly gentleman named Malcolm Lloyd, who was bearing much of the expense of the expedition and who was eagerly anticipating the opportunity to participate in the work; a young student named Ned Page; a local Nebraska boy, John House, whom we hired to cook for us; and myself. The four of us from the East went out together on the train (planes were still a somewhat unusual way to travel), rented a small truck at Scottsbluff, Nebraska, and made our way across the grassy plains to the Cook Ranch, officially designated as Agate because it had a post office. (The post office, housed in a tiny building by the side of the road, was presided over by Captain Cook's brother, Uncle Jack, a character of the first water.)

Harold Cook was on hand, and we went with him to the two buttes, where he showed us a spot that he thought would be productive if we were to dig there. This quarry was on the side of the larger butte, Carnegie Hill, so named because the Carnegie Museum had made the first large excavations here, with the enthusiastic support of its prime patron, Andrew Carnegie. The other hill was called University Hill, since the University of Nebraska had made the initial excavations there.

We made camp at a little cabin provided for bonediggers by the Cook family. It was near the base of Carnegie Hill and was surrounded by a barbed-wire fence to keep the Cook cattle at a respectful distance. There was a pump by the cabin, and cold, sweet water flowed from it. Beneath us lived a family of skunks. The cabin was our cook shack and dining room, but we slept in tents that we set up in the fenced enclosure. It was a comfortable camp.

Every day we worked at our quarry; every evening we could sit outside our cabin and enjoy the spacious quiet of the vast, rolling plain. On Saturday afternoons we would drive into Scottsbluff and buy groceries for the coming week, after which we would take in a movie. Then a late-night visit to an ice plant where we would buy a 200-pound cake of ice, back through the cool of the night to our camp, and as soon as we arrived we would bury the ice in a pit filled with sawdust, specially prepared for it. All week John would keep the refrigerator in our little cabin well stocked with ice, and so we had fresh butter, meat, milk and cream, and cold drinks.

To begin our work we dug a large cavelike excavation in the vertical wall of Carnegie Hill. The bottom of this cave was just above the bone layer. Then with the usual complement of awls and small picks, whisk brooms and brushes, shellac and tissue paper, we exposed and protected the bones that came to light.

Various ancient mammals had been interred in this deposit, but the most common one was the little rhinoceros, *Diceratherium,* distinguished by the possession of side-by-side horns on its nose. Associated with this rhinoceros was *Moropus,* the large horselike chalicothere having large claws on its three-toed feet instead of hoofs. Another member of the fauna was the giant piglike entelodont, *Dinohyus.* (It may be remembered that during my apprenticeship years at the University of Nebraska I, with expert help from Henry Reider, had set up skeletons of these three beasts. So I was now digging for old friends.) The bones gradually came to light as we worked in our quarry, and in time we had a large area of bones uncovered—most of them the remains of the little rhinoceros, but also with *Moropus* and *Dinohyus* present.

After many days of work there was a huge block, full of bones, ready to be plastered and undercut. Plastering the top and sides of the block was routine, but the problem of undercutting it was formidable. We tried using an old logger's saw but the sandstone of the block quickly wore away the teeth. We also tried the trick of pulling a piece of barbed wire back and forth, with one person on each end of the wire, but that didn't work; the barbs were quickly bent or pulled out of place. In the end it was a matter of long days whacking away with picks.

(Things today are more up to date. Gasoline-powered rock saws have been developed that do such work quickly and with much less expenditure of muscle power.)

It will be readily appreciated that we could not completely undercut the block; such a procedure would at some moment have brought the huge mass, weighing several tons, crashing down on the person who might be half underneath it. So we left the block supported on several rock pedestals; in effect we had dug a series of broad tunnels beneath it. Through these tunnels we thrust heavy timbers, to form a cradle beneath the block, while other timbers of equal weight were bolted to the ends of these bottom units, and were extended up the sides of the block, to be bolted in turn to timbers across the top of the block. Thus the block was contained within a heavy frame, securely bolted together.

Now it was ready for removal. We went into Scottsbluff, located a contractor, and made arrangements for him to come out with a big truck equipped with a powerful winch. In due time he came, cables were attached to the block, and the winch went to work. Slowly the heavy mass of rock, enclosed in its plaster jacket and surrounded by its timber frame, was eased up onto the bed of the truck. After that it was a matter of hauling the block into town, to the freight station. Our summer was over.

What has been said in these last few paragraphs undoubtedly is prosaic. But the collecting of fossils can be prosaic, especially if work is being carried on, day after day, at a quarry. It is all a matter of patience, much application, and plaster in unbelievable quantities.

The less prosaic part has to do with uncovering the bones and trying to identify and analyze the things that are being found. This can be truly exciting. One brings to light a vertebra, then another, and then another. Will they lead to a skull? That is the fond hope—sometimes realized, often not. Will there be legs and feet? And what does one do if the animal is strange—something completely new to science? That is the real thrill. Sometimes the work seems all too slow, and everyone in the quarry is consumed with impatience to see just what this new animal is like. Yet patience is the watchword; one cannot make the mistake of trying to work too rapidly, thereby running the risk of destroying some crucial part. The specimen must be as perfect as possible, in order that it can be properly studied, after it has been finally cleaned in the laboratory.

These are the quiet compensations for the daily work in the quarry, where the sun is hot and thirst is ever present. One can forget discomfort when seeing a fossil skeleton come into view bone by bone, a skeleton that has been entombed in the dark rock for many millions of years.

We had such compensations and pleasures at the Agate Springs Quarry, and others, too. How can I put into words the enchantment of the early mornings, when the sun was low and the lark buntings flew in hurried flocks across the waving grass of the prairie? Or the excitement one morning when we drove up to the quarry and five graceful antelope leaped up from where they had been resting by our fossil bones, to pause for a moment on the skyline before they drifted down the opposite slope to lose themselves in the tall grasses and the eroded breaks of these high plains? Or the calm of the evenings when great clouds lifted high into the vast sky, to be bathed in glorious colors by the setting sun? Nowhere, I think, are there clouds and sunsets like those to be seen on the western plains.

Every evening when we had completed our work on the bone slab we would cover it with a tarpaulin, to protect the delicate, exposed bone surfaces, as yet not covered by plaster jackets, against a possible rain, or the probing nose of a coyote. And every morning when we rolled back the canvas to start our day of work there would be a little white-footed mouse, who during the night had pulled some strands out of our stack of burlap sacks to begin the construction of a nest. Off she would scoot, and we, unfeeling creatures that we were, would throw her little nest aside, because it was always right in the middle of the area where we had to work. The performance went on morning after morning, until finally, I suppose, she decided that there must be better places to rear a family. We saw her no more.

I said something above about the quiet of the evenings, but they were not completely quiet, because a couple of miles away, on the crest of a ridge, was an oil well hopefully going down. We could hear the clank of the machinery, muted by the distance, and at night the lights on the rig shone like a multifaceted beacon through the dark sky. (The work went on, night and day, but in the end I don't think any oil was found.)

Late every afternoon the man superintending the drilling would fly in, piloting his own plane, and land at the oil well rig. Then, just about the time we were finishing supper in our cabin we would hear the roar of the plane approaching and rush out to wave at him as he skimmed the top of our cabin with only a few feet to spare. It was our way of saying a farewell to each other for the evening, and I know he enjoyed it.

Finally, one of the very special joys of working at the Agate Quarry was the time spent during off hours at the Cook ranch house in its grove of trees. There was a luxurious pleasure in resting on the grassy lawns during a dreamy Sunday afternoon, enjoying good talk and the cool environment of this western oasis. There was much good fellowship in the Cook drawing room, where Margaret Cook, an accomplished musician, would play the piano, and where at times we would all join in song. Indeed there is much to be remembered of the summer at Agate, and not all of it had to do with the collecting of fossils.

The year of the Agate Springs Quarry adventure was followed by that dark summer of 1939, when in Europe the unthinkable became reality, and a part of the world took fire as the savage war machine of the Third Reich crashed through Poland. We listened to the broadcasts with dread and apprehension and we read the papers with downcast hearts. What was it all to mean—to the world in general and to us in particular?

For the moment there was no real impact on our lives. That was the summer that our second son, David, was born, and of course this was an event to occupy our thoughts and our time. During the following summer our third son, Philip, was born, and we were more than busy at our home. I continued studies at the museum on Asiatic mammals.

Then in the spring of 1941 plans began to take shape for a collecting trip, this time to the famous White River Badlands of South Dakota.

The White River Badlands, a large area of fantastic erosion in Oligocene sediments, immediately to the southeast of the Black Hills, has long been a prime hunting ground for students of fossil mammals. Here on either side of the White River, which flows eastward to the great Missouri, are white and pastel-colored walls and pinnacles, eroded buttes and tablelands, rising starkly several hundred feet above flood plains that in part are barren stretches of sand and clay and in part are grassy floors where white-faced cattle graze. In these ancient sediments, blinding white and hot under the summer sun, are the fossil bones of mammals that long ago vanished from the American scene. Some of these mammals would seem reasonably familiar to our eyes—little insectivores and rodents and rabbits. Others would seem very strange: small three-toed horses no larger than whippets; graceful, fast-running rhinoceroses; other rhinoceroses that lived in the rivers like modern hippos; gigantic horned browsers called titanotheres, superficially like tremendous rhinoceroses in appearance; little camels; antelopelike animals abundantly supplied with six horns on the skull; primitive clumsy carnivores; saber-toothed cats, and many others.

These animals were first made known to science during the middle of the nineteenth century, particularly by the pioneer paleontologist Joseph Leidy of Philadelphia. Since Leidy's time the White River beds have been explored time and time again by many institutions, and large collections of those Oligocene mammals have been made. The fossils are abundant, erosion is rapid, and thus new materials are constantly coming to light, to be lost to the forces of erosion if not immediately collected.

The American Museum specimens of White River mammals had become much depleted by exchanges with and gifts to other museums, so in the spring of 1941 Walter Granger decided that it would be an excellent idea for us to go out to the White River country to renew our collections. Plans were made. The area was by then a National Monument, and protected, so the museum obtained permits from the government to collect fossils on the federal land. I was to accompany Albert Thomson, a veteran paleontological technician and collector. Thomson had joined the museum back in the nineties, as had Granger, and they had been close friends for many decades. Thomson, known to all of us as "Bill" (for what reason I do not know), and I would begin the work, and Granger would join us later in the summer.

Here again, as in the case of the work at Agate, fieldwork in the White River Badlands would not be directly in line with my research program. But it was a museum effort that needed to be carried out that summer, and I was a logical person to help.

Bill Thomson and I made our way by train to Rapid City, South Dakota, from which place we would equip ourselves for an extended stay in the badlands. A museum car was on hand for us; we spent a few days getting equipment assembled and buying groceries, and then we drove south and east to the little settlement of Scenic.

The town of Scenic belied its name. There were some spectacular badlands outside town and at some distance, but the town itself was a discouraged place. It had been settled originally as a little shopping center and shipping point on a branch of the Milwaukee Railway. But with the advent of the automobile the importance of the town gradually diminished as the local ranchers deserted Scenic to drive to Rapid City for their goods. When we were there Scenic consisted of a "hotel" of sorts, a bar next to it, a store across the dirt street, a small restaurant down the line from the store, and a gas station close to the railroad track. There were a few houses scattered here and there, and off to one side a little Catholic church with its attendant dwelling for the priest.

The priest, Father Balfe, was a charming and a very literate man. I think he had been consigned to Scenic by an unsympathetic bishop, but he was making the best of a discouraging situation. He was particularly interested in our work and was a good companion. Our other close friends in Scenic were Claude and Celia Barry, who ran the store. They, too, were interested in our work and they gave us encouragement and help, including one amenity that I greatly appreciated. This was a semi-detached outdoor shower, provided with water heated by the torrid summer sun. It *was* torrid; I recall that during one solid month the temperature every day was above 100 degrees. It was a real pleasure and a refresher to come in after a hot, debilitating day of fossil hunting and enjoy a good shower before the evening meal.

When we arrived in Scenic, Bill and I established ourselves in the hotel, where our quarters consisted of a room occupying one wing of the little building, with a separate door opening on to a dirt road that ran between the hotel and the railroad station. We were quite shut off from the rest of the building, which gave us a feeling of privacy. Alas! Any feeling of privacy was for me dispelled during our first night in our lodgings, because I soon found that our room was occupied by a considerable population of *Cimex,* the common garden variety of bedbug. The bugs didn't bother Bill, but they were too much for me, so the next morning I moved out of the room, to make my quarters in our field car, which at night we parked in front of our room. For the rest of the summer that was my home, and it turned out to be very comfortable indeed. It was a sort of panel truck, and every evening I would spread my sleeping bag in there, close the back doors, open the windows by the front seats, and spend a truly private night, delightfully cool and unmolested by insects. Generally just as I was falling asleep I would hear the train come through on its way to Rapid City. Bill stayed in the room; how he put up with it, I don't know.

Our work in the badlands was very different from the experience at Agate, three years previously. There we had gone to a predetermined spot, where we spent all of our time working in the quarry to remove a huge block of fossils. In the White River Badlands we were *searching* for fossils and removing the significant ones that we found. We spent much of our time just looking, ranging back and forth across the barren flats with our eyes constantly on the ground or following the contours of the badland hills, scanning their slopes up and down for the telltale signs of fossil bones. It could be tiring, and in a way boring, wandering across the landscape hour after hour without finding anything of consequence. Or it could be intensely exciting, especially when a few scraps of bone would be the clue to a beautiful and magnificent specimen. Then the search would be halted, and our energies would be directed at getting the fossil out of the ground, using the same methods that have been described, except that here we were dealing with objects much smaller than the Agate block so our logistical problems were simpler.

When the specimen was removed, we would take up again our interrupted search for more specimens. It was a matter of spotting the fossil in the ground—by its shape and by its texture. These are the clues that distinguish a fossil from the surrounding rock. Shape is important, but texture is especially significant. The nature of the fossil usually reveals its organic structure—the distinctive surface of bone, the intricacies of teeth, the rugosities of scales. Such features are readily learned, and once learned are not forgotten. Thus the brilliant sunlight reflecting from the surface of a fossil bone or from a tooth would reveal the treasure to a knowing eye. One could not hurry; the ground must be examined thoroughly. And always it was hot, brutally hot.

Bill Thomson in the White River badlands. We spent much of our time just looking, ranging back and forth across the barren flats with our eyes constantly searching the ground.

Bill Thomson and I usually prospected the ground separately but within sight of each other, thereby covering the terrain more efficiently than if we had walked side by side. We would slowly advance on parallel lines, perhaps a hundred yards apart, perhaps as much as a quarter of a mile distant each from the other. As we walked along we alternately looked at the ground around us and glanced up toward each other, like a couple of terrestrial vultures keeping a lookout for the one who first would spot a carcass. If one of us found a fossil and sat down to probe it, the other would soon come over to see what it might be, and to help if necessary. There was always a feeling of friendly rivalry, each member of the team hoping to find something that would bring the other partner over to share in the discovery.

At times we did become separated, especially if we were working among the badland hills. In such cases we had a predetermined rendezvous, generally at the field car, which would be parked at a strategic spot. That was where we would eat our lunch, and if we had been separated for an hour or two or three we would then compare notes and any small specimens picked up, and make plans for the afternoon.

One of my memorable experiences was the time when I was exploring by myself and was joined by a herd of wild horses. They seemed to be very much interested in what I was doing and followed me as I walked along, or stood around in a respectful semicircle when I would sit down to probe a prospect. This went on for an hour or so, and little did they realize that I might at any moment find the remains of one of their very distant three-toed ancestors. Finally they had had enough; away they went with tossing manes past the nearby badland hills and on to some grassy flats in the distance.

The most exciting and pleasurable event of the summer for me was the day I discovered a skeleton of *Hyaenodon,* one of the early carnivorous mammals known as creodonts that inhabited western North America many millions of years ago. *Hyaenodon* was a large creodont, about the size of a wolf but of heavy proportions, with a robust skull and strong crushing and cutting teeth. It was one of the last of the creodonts—these rather clumsy carnivores had been largely replaced by modern carnivores such as dogs, mustelids, cats, and their relatives. It had been a long and discouraging morning, and a little before lunchtime I clambered up to the top of a small pinnacle, in part to have a look around the countryside and in part to get the benefit of any cooling breeze that might be available up there. As I was looking around I glanced down and there below me, on a little sort of eroded terrace on the side of the hill, I saw a skeleton. I could hardly believe my eyes, for it obviously was not the skeleton of a sheep or a coyote or any other modern animal. It was a fossil, and it was nicely articulated and almost all there, which was unusual because fossil skeletons usually don't re-

Beneath the welcome tarpaulin we worked on the skeleton for several days.

main intact once they are exposed.

With an inward if not an outward whoop I almost flew down the slope in long bounds to get a close look. It was a beautiful skeleton of *Hyaenodon,* stretched out on its side just as the animal had died some 40 million years ago. I hurried back to the car, where Bill was waiting for me, and in a state of euphoria I told him about the discovery. We had our lunch, and then we both went back to the site to appraise the specimen. No doubt about it—we would have to spend several days there collecting it.

The next morning we drove to the base of the hill and then hauled our collecting equipment up to where the skeleton was exposed. Bill started the preliminary work of cleaning around the skeleton and applying shellac to harden it, while I took a long look at the rising sun, already white hot in the eastern sky.

"Let's put up a sun shade!" said I.

"Oh, why bother," said Bill, who already was beginning to sweat profusely.

"All right," I said, "you can sit there and cook if you like, but I'm going to rig up a shade."

Bill didn't object, so I set to work. I found some poles and hauled a tarpaulin out of the truck. It took a half hour or so, but in the end I had a sun shade arranged, much to my satisfaction and to Bill's also, for it was quite apparent that he was grateful for the protection it offered. In that dry climate one could be reasonably comfortable in the shade, even on the most scorching days. Beneath the welcome cover we worked on the skeleton for several days, hardening the bones, isolating our block, and finally encasing it in its bandage and turning it over. The specimen is now on display at the American Museum of Natural History.

In the midst of this torrid collecting work we had a bit of a respite; Margaret came through South Dakota with her mother and her stepfather, Ralph Minor (professor of physics at the University of California). Some kind friends (kind indeed!) had taken over the care of our three small boys, so she was enjoying a drive across country for a much-needed vacation. Since her mother and Bill Thomson were old friends, we had a get-together in the badlands, and then we all went to Rapid City to cool off and rest for a day. Our vacation was all too short. Margaret and the Minors went on west, and Bill and I returned to the badlands.

Our work continued. In late August, Walter Granger came to join us. He was due to arrive one evening on the Milwaukee train that passed through Scenic, and we planned to give him a rousing welcome when he stepped down from the train. It was a noisy greeting, even more enthusiastic than had been foreseen, because on the very afternoon before he arrived Harold and Margaret Cook appeared in Scenic, and with their connivance we staged a hilarious meeting, complete with music on a guitar by Margaret and a lei of onions that we draped over Granger's neck when he alighted. Father Balfe and the Barrys joined us in the fun, and it was fun. I will never forget Granger's roars of laughter and the rather startled looks of some of the passengers on the train as they peered out the windows, trying to make out through the darkening sky just what in the world was going on.

For a few short weeks we enjoyed the companionship of Walter Granger in the field during the days and in Scenic during the evenings. Then by prearrangement we made a trip through the Black Hills and from there down to the Cook Ranch at Agate. The Society of Vertebrate Paleontology had scheduled a field conference, which I would attend. The conference assembled in northwestern Nebraska, so I found a berth in one of the cars of the conference caravan, leaving Granger and Thomson at Agate, where they would rest and visit with the Cook family.

We had an interesting, peripatetic meeting lasting about a week, during which time we visited various fossil localities, finally ending up at the University of Nebraska in Lincoln, where there was a banquet. Then I made my way to Lusk, Wyoming, by train and by bus, where I was to join Granger and Thomson.

As I stepped down from the bus I was met by Bill and Harold Cook, both of them very quiet and sad. Walter Granger had died in his sleep the night before. He was sixty-eight years old; his death was caused by a heart attack.

I was stunned. Never before had death come so close to me; it had a devastating effect. Walter Granger, so great a friend and my revered mentor, was gone. That day I was numb, and to distract our minds as much as possible from the unexpected event Harold Cook drove Bill and myself on a little trip around some of the nearby Wyoming oil fields.

Then Bill and I went back to Scenic to collect our gear and to bring our planned fieldwork to an abrupt halt. As we entered the dusty little settlement the early September wind scattered sand before us and rattled the few signs on the short main street. The dry wind seemed to sound a dreary requiem that brought tears to my eyes.

That was the end of the summer; Bill and I returned to New York by train, carrying Walter Granger's ashes with us, and at Pennsylvania Station we were met by Roy Chapman Andrews, the director of the museum and a close friend of Granger, by other museum friends, and by Margaret. I went home with her to rejoin our family, and two or three days later she arranged for someone to stay with our three little boys while we took the day off and went to the Bronx Zoo. That also helped to occupy my mind and to ease a bit the sense of loss that kept recurring many times each day.

The autumn passed, and then the war came—on that fateful Sunday of December 7. I need not go into the details of how it changed our lives; it affected everybody in the country. For me there was no immediate, sudden change, nor was I eventually to be called into service. On two occasions I thought that I might join the armed services; once when the Army asked me to apply for a commission because it wanted officers who were acquainted with deserts, and once when the Navy made a similar request because it wanted officers to collect and study mammals that might be vectors of tropical diseases in the South Pacific. But on both occasions I was passed over, perhaps because they found other candidates who did not have small children dependent upon them. George Simpson left our department in the museum to serve in Africa. Bobb Schaeffer, who was to join the department immediately after the war, also served in Africa.

I was left behind, and then there came a change of direction for me—a change in no way connected with the war, but rather the result of happenings at the museum.

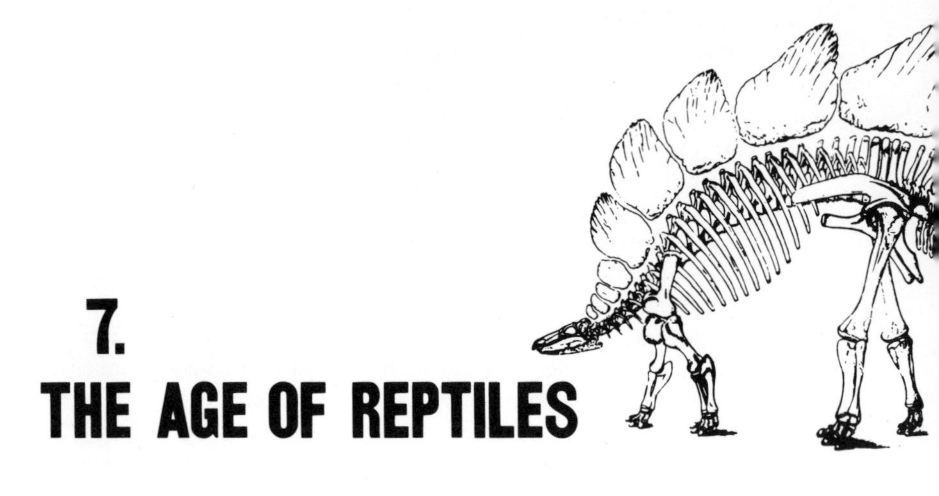

7.
THE AGE OF REPTILES

In 1942 Barnum Brown retired after many decades of service as Curator of Fossil Reptiles, distinguished by his collecting an unparalleled series of dinosaurs, and I was appointed to succeed him. Although I had up to this point done little if any research on fossil reptiles, I was and had been for some years much interested in these ancient vertebrates. I entered upon my new career with hopeful anticipation.

It *was* a new career involving a shift of interest and emphasis from the middle and late Cenozoic mammals, upon which I had until now focused my attention, to the "lower tetrapods," the amphibians and reptiles, and particularly those that lived prior to the dominance of the mammals which began about 60 million years ago. Perhaps a little diagram will help.

CENOZOIC ERA	Pleistocene Period	The Ice Age	
	Tertiary Period Pliocene Epoch Miocene Epoch Oligocene Epoch Eocene Epoch Paleocene Epoch	The Age of Mammals 65 million years ago	Nebraska apprenticeship The restricted years Studies of Asiatic fossil mammals
MESOZOIC ERA	Cretaceous Period Jurassic Period Triassic Period	The Age of Reptiles 225 million years ago	The change to lower tetrapods, with eventual emphasis on the Triassic
PALEOZOIC ERA	Permian Period Carboniferous Period Devonian Period Silurian Period Ordovician Period Cambrian Period	The Age of Fishes and Invertebrates 600 million years ago	

There are various ways in which to define one's research interests. A common approach is to pick out or to get involved with some problem or some related set of problems, and to stick with this interest from beginning to end. Another more pragmatic approach, and this was the one I followed, is to maintain a flexible attitude and to make use of opportunities as they arise. Here was an opportunity that I felt could not be ignored. Moreover, I felt that even though my work had for a decade and more been concentrated on mammals, I could make the shift to amphibians and reptiles thanks in part to the broad and thorough training I received from Professor Gregory. It was a big change and it would have to take place over a period of time. I did continue some work on mammals, especially to complete projects that were under way or scheduled, while at the same time I began studies on ancient amphibians and reptiles.

The American Museum of Natural History has a superb collection of all fossil vertebrates, and the bones of amphibians and reptiles form an important part of this collection. Thanks to work initiated by Osborn when he first came to the museum in 1891 and carried on by Barnum Brown, there is an outstanding series of dinosaurs at the museum, some of them exhibited in the two spacious dinosaur halls, and many housed on gigantic steel racks in the storerooms. These are the reptiles that catch the public eye. Yet the museum has important collections of other lower tetrapods, especially Permian and Triassic amphibians and reptiles from North America and Africa.

I liked the dinosaurs, and indeed I was in the course of time to write some books about them, but I was especially interested in the earlier tetrapods, the strange fossil amphibians and the varied reptiles of the Permian and Triassic periods, among which latter are included ancestors of the dinosaurs. Consequently, I decided to attempt studies on these earlier backboned animals, for I saw various important problems that could be pursued advantageously. One of the nice things about paleontology is that there are always research problems waiting to be done. New fossils are always being found, and these frequently change our concepts about whole groups of animals—about their adaptations and relationships and about their distributions. Moreover, the occurrences of fossils have crucial bearings upon past continental relationships and earth history, as subsequently we shall see. I therefore began to review the field, to see just where I might begin.

The war was on in full fury, with one consequence that such esoteric activities as vertebrate paleontology, and basic research in general, were of even lesser import than was usually the case. I carried on as best I could under the conditions that prevailed. Then, in the autumn of 1943, something happened.

The assistant director of the museum, Wayne Faunce, called me into his office one day and started to talk about a field car that had been stored in Glen Rose, Texas. It had been used by Barnum Brown and R. T. Bird, Brown's talented field assistant, during the summer just before Pearl Harbor. R. T. (as we all called him) had stored it with the intention of picking it up the following summer for another field campaign, but with our entry into the war that plan had been cancelled. What were we going to do, Mr. Faunce inquired. The car, which was filled with equipment, was running up storage charges.

I did a bit of quick thinking and suggested that I go to Texas, pick up the car, and drive it back to New York, a plan that was immediately approved. Then I got in touch with Professor Alfred S. Romer of Harvard, to inform him of the situation.

Al Romer was then, and for many years after, the leader among American paleontologists in the study of ancient reptiles and amphibians; indeed, he was a respected world authority on this subject. He was a vertebrate paleontologist of vast erudition and an outstanding figure in the world of science. (Years after the war he was elected president of the

Al Romer was then, and for many years after, the leader among American paleontologists in the study of ancient amphibians and reptiles.

American Association for the Advancement of Science.) Much of his research was concentrated on the tetrapods found in the Permian red beds of Texas.

I got a letter off to Al, proposing that we both go to Texas, pick up the car, and on the way back stop and do some exploring in the Texas red beds, which were right on our way. This was important, because in those days of very strict gas rationing we would be allowed only enough fuel to get the car back to its home base. Al was delighted with the prospect; it would give him the opportunity to check on his favorite fossil beds, an opportunity otherwise not available because of travel restrictions. It would give me the opportunity to spend some time in the field with a great scientist—to pick his brains and learn a thing or two. We made plans.

We met in Dallas; Al was accompanied by Henry Seton, a Harvard worthy whom I had known for quite a number of years. We took a bus to Glen Rose, and then went to look at the field car. The car, with its load of equipment, had been sitting idle on its tires for more than two years. The tires were as flat as pancakes.

I will not attempt to describe some of our trials and tribulations. We had a government order for new tires, but how were tires to be obtained? Fortunately, the garage man in Glen Rose had to make a trip to Dallas and by rare good luck he found a set of tires just the size we needed. The tires were put on the truck, the engine was overhauled and tuned up, and we were ready to start.

It was a cranky old car, but it got us there. For about three weeks we poked around in the Texas Permian, studying the sequence of the strata and looking for fossils. Henry stayed with us for a while, but then he found the weather a bit too cool for his liking so he returned to Cambridge. From then on it was just Al and myself.

A more delightful field trip I have never had. Al had an ebullient sense of humor, and he kept me amused and often in stitches during our entire trip. Furthermore, just as I had hoped, he talked at length about primitive tetrapods, thereby giving me an unparalleled introduction into my new area of research. We did not attempt any large-scale collecting but we did pick up some nice things, so the trip was paleontologically productive in a small sort of way. (Years later I learned, much to my gratification, that Al, too, had considered this one of his more pleasant field trips.)

We got the car back, not without some problems to the two of us. Because it had been flat on its springs for a couple of years, the springs had no life to them. So the old truck, which we had christened "Geraldine" after one of Al's favorite fossil localities, gave us a spine-jolting ride all the way home from Texas. The result was that Al and I had bad backs for years afterward, and I still have a somewhat fragile lumbar region, owing in part perhaps to that hard-riding old car.

Then came 1944, a year memorable to us because of the birth of our fourth son, Daniel, in March. Early during that year I had been interested in reading a paper by Professor Raymond B. Cowles of the University of California in Los Angeles having to do with the extinction of the dinosaurs. Cowles was proposing that the dinosaurs may have become extinct because temperatures at the end of Cretaceous time—the end of the "Age of Reptiles"—rose to such extremes that the ruling reptiles could not survive. This was just the opposite of a prevalent view, namely that there was a worldwide cooling of temperatures which may have had a bearing on the disappearance of the dinosaurs.

I wrote to Cowles and he wrote back, and we had quite an exchange of letters. The whole issue revolved around the problem of temperature tolerances in large reptiles: how much heat and how much cold could they stand? Why not get together and work on the problem?

Cowles had been the major professor under whom Charles M. Bogert, the Curator of Herpetology at the American Museum, had done his graduate work. The American Museum had a field research station in Florida, and plans were made for Chuck Bogert, Ray Cowles, and me to go to Florida that summer, to experiment with alligators. Why alligators? Because they are large reptiles and the living reptiles most closely related to the long-extinct dinosaurs. It seemed like a very good idea.

Ray came to New York, where I became personally acquainted with him, and then Ray, Chuck, and I went by train to Florida. We made a most congenial team. Chuck had for years been one of my closest museum friends, and Ray soon became another valued friend—from then until the end of his life. He was truly an interesting person, an original thinker. His had been a life of more than ordinary aspect. Born in southern Africa, he had grown up among the Zulus, whose language he spoke fluently. It was fun to listen to Ray talk Zulu, one of the click languages. During his boyhood years Africa was still a land largely unspoiled by the inroads of modern "civilization," a land of many animals. Ray had many African stories to tell.

We were to spend a month or so at the Archbold Biological Station, not far from Sebring, Florida. This was a reserve of a thousand acres or so which had been purchased by Richard Archbold, a friend of the American Museum, and maintained by him for the use of visiting scientists. Dick lived at the station, where he made life interesting in many ways. He was all kindness and more than ready to help us in any possible way.

Dick had one handicap—he was plagued by a physiological condition which caused him to drift into sleep at any time, at any place. For this reason he had been rejected for service by the army. At dinner in the evening Dick would suddenly go to sleep; he would sleep for four or five minutes and then as suddenly wake up, to resume his end of the conversation where

Ray Cowles had many African stories to tell.

he had left off. He did love to drive a car, and a ride with him could be a spine-tingling event. There was an old family retainer at the station, an Englishman by the name of Mr. Walters, and he always made it a point to jump into the car next to Dick whenever Dick took the wheel. Mr. Walters would grab the controls if Dick went to sleep. Furthermore we all managed to talk in a very animated fashion when we rode with Dick.

The Archbold Station was a comfortable place, with solid buildings, designed to be hurricane proof. There were nice living quarters and well-equipped laboratories, and lots of room for outside work. After getting ourselves settled we were ready to start the project, except for the fact that we had no alligators. Chuck had made arrangements for a graded series of small to large alligators to be supplied by Ross Allen, the proprietor of the Allen Herpetological Institute, but these were not immediately at hand. For several days we had to do other things—mainly study the local reptiles, of which there was an abundant supply.

Then one evening a large male alligator walked right into camp, and we greeted him with delight. He was just what we needed for the upper end of our series. Shortly thereafter the other alligators arrived, so we had the necessary reptiles, ready for our experiments.

We kept the alligators in a series of pens at the station; the big male was segregated in a large fenced enclosure with a pool in the middle of it. There he was monarch of his domain, a fact that he did not hesitate to impress upon any person or animal who approached his little kingdom. If we went out at night to check on him we would first see the red glow of his eyes in the light of the electric torch. There would be a series of loud hisses and a charge, and he would hurl himself at the fence with all of his might. He was a hostile and formidable saurian, not to be trifled with.

The first time we tried to capture him for use in our experiments we attempted the technique of carefully placing around his neck a rope noose at the end of a long bamboo pole. He glared balefully at us while Chuck proceeded very slowly to get the noose in place. In a flash he grabbed the bamboo pole in his viselike jaws and with lightninglike rapidity started whirling over and over in the water. Spray was thrown high in the air as the alligator reduced his end of the pole to splinters, thus tearing one end of the rope noose loose from its attachment. Chuck managed to hold on to his end of the pole while Ray and I retreated in trepidation from the angry reptile.

Our next strategy was to capture him on dry land. One of us would approach him from the front to engage his attention. The second member of our trio would then circle around and grab his tail, holding on for dear life, while the third player on our team would jump on his back to hold him down. Thereafter it was a matter of tying his jaws shut with a piece of stout rope. This was not as difficult as it may seem at first glance. Crocodiles and alligators have extremely powerful jaws and cruel teeth, as long as they are closing the mouth. But the apparatus for opening the mouth, the *depressor mandibulae* muscle, is relatively weak. One can push the alligator's head down against the ground, get hands around his jaws, and rather easily hold them shut. That was what we did; one of us would hold the jaws together while another would tie them shut with the rope.

These techniques were used on the smaller alligators too, with greater ease and less risk. But the risk was always there even with the little fellows, as Chuck found out one day when he was snagged on the hand by an alert reptile. The blood flowed freely for some time.

We subjected those unwilling saurians to a whole series of experiments; exposure to sunlight in different poses, confinement in a constant-temperature chamber, exposure in tanks to increasingly warmer water, and conversely to increasingly cooler water, as well as other tests. The alligators must have become pretty fed up with the whole business before it was over. Nevertheless, their reluctant cooperation taught us many things about temperature tolerances in crocodilians, with interesting implications concerning their ancient relatives, the dinosaurs. And at the end of the month they were returned to their usual habitat, perhaps somewhat annoyed by the experience but otherwise none the worse.

We learned that the larger alligators when exposed to the sun or to other sources of heat showed a slower temperature rise than the smaller alligators, because it takes longer to heat up a large mass than a small one. Likewise, cooling was slower in the large reptiles than in the small ones.

Charles Bogert contemplating our guest, a large male alligator who had walked right into camp. We greeted him with delight.

(Modern reptiles are "cold-blooded" ectotherms, which means that they have no internal mechanism for temperature control. Their temperatures are determined largely by external environmental conditions.) We were able to determine the rates of temperature increases and decreases as they are related to size.

The implications were obvious. If the dinosaurs were ectothermic, as are their modern crocodilian cousins, then the rates of bodily increases and decreases of temperature in the giant forms must have been extremely slow. Therefore, these ancient giants, instead of having rapidly fluctuating body temperatures as do most modern reptiles, may have had rather constant temperatures, thus enjoying some of the advantages of endothermic, or "warm-blooded" animals (like ourselves). Of course if the dinosaurs were independently warm-blooded, as is now maintained by some authors, different factors must be considered. However, the purported endothermism of dinosaurs can not be definitely proved; it is an inferred physiological trait, the inference being based upon some rather esoteric features of bony anatomy and conditions of fossilization. In brief, it is an open question.

Whether we found out anything about the extinction of dinosaurs is doubtful. But I think we did get some insight into dinosaurian physiology and how these ancient reptiles may have behaved. At any rate, we published a paper on the Florida alligators in the scientific series of the American Museum, and I had the satisfaction, years later, of seeing this referred to as a "seminal contribution" in its field.

That was our Florida adventure, except for a few supplementary activities unrelated to alligators. One of these happened on a day when we went down the Caloosahatchee River in a motor launch. Along the way we saw something sticking up out of the water and we stopped to investigate. Lo and behold, it was a fossil leg bone of a mastodon, one of the extinct proboscideans or elephant relatives that roamed across North America in such numbers during the last Ice Age. The water was only about waist deep so we waded out and started probing in the mud with hands and feet. The upshot was that we found a number of mastodon bones, the remains of an animal that had lived in Florida many thousands of years ago.

In the following year of 1945 the war was drawing to its climactic ending and scientific activities were still much reduced. That summer Chuck and I drove across the continent to California, where I was to teach a summer course in paleontology at the University of California in Berkeley and where Chuck was to teach a course in herpetology at the University in Los Angeles. Wartime driving was restricted and we were rationed just enough fuel to make the trip—at a speed of thirty-five miles per hour. We loafed across the land and it was one of the most enjoyable drives I have ever had. We knew we could not hurry and thus we could enjoy the countryside as we drove along the all but empty roads.

It was in Berkeley that summer that Margaret's aunt, Lee Thayer, a detective-story writer, composed a murder mystery based on our Florida work, and I helped her with it. It was a lot of fun.

I was in California when V-J day was announced over the radio, and the crowds went wild. That was at the end of the summer, and soon I took the train for the east coast. On the way I stopped in Arizona to look at Triassic rocks in the Painted Desert and the Petrified Forest. By now I had settled on the Triassic as the area for my research in the immediate future, and I wished to have a preliminary survey of the sediments in Arizona preparatory to my coming out during the following summer to do some serious collecting.

8.

THE LITTLE DINOSAURS OF GHOST RANCH

Why the Triassic? Because, it seemed to me, the Triassic period of earth history was a very crucial time in the evolution of the backboned animals. It was a period of transition, when the older lines of tetrapods (these being the land-living vertebrates) were giving way to new amphibians and reptiles which were to occupy the continents. Some of the amphibians and reptiles that had developed so successfully through the preceding Permian period lingered on. Such were certain large labyrinthodont amphibians, animals often six feet or more in length; such were a few of the primitive reptiles known as cotylosaurs, the Triassic representatives being small, inoffensive, "lizardlike" forms; such were the therapsids or mammal-like reptiles, certain genera of which were trending toward the threshold separating reptiles from mammals. But these persistent remnants were being overwhelmed and displaced by a horde of new tetrapods—the earliest frogs, and the earliest turtles, and the earliest lizards, progressive reptiles known as thecodonts which were to be the ancestors of the ruling reptiles of Mesozoic times, and the stem members of these thecodont descendants, the crocodilians and especially the dinosaurs. The first dinosaurs, appearing in late Triassic time, founded evolutionary lines that were to proliferate and dominate the continents for 100 million years. That in itself makes the Triassic a decisive time in the history of life. Finally, the Triassic was the period when the first mammals made their appearance, derived from some of the mammal-like reptiles. These first mammals were very small and they seem insignificant, but they were the progenitors of animals that were destined to succeed the dinosaurs as the rulers of the land. But the flowering of the mammals was not to take place until long after the close of Triassic time. The Triassic period belonged, above all, to the reptiles.

It was my plan to carry on a Triassic campaign in the Southwest, at least for several years. There are extensive Triassic exposures in Arizona and New Mexico, and it seemed to me that these were ready for exploration. There had been no significant paleontological work in these Triassic beds for several years, and the forces of erosion must undoubtedly have exposed fossils that hitherto had not been visible. Indeed, Dr. Charles Camp of the University of California, who in previous years had worked in the Chinle Formation of Triassic age, had shown me around some of the promising areas in the summer of 1945 as I was returning to the East from my stint of teaching in California, and had advised me about various localities.

Not that the Southwest is the only place in North America to do Triassic work. There are wide exposures in Texas, in Wyoming, and throughout the eastern coastal states and the Maritime Provinces of Canada. Some very interesting Triassic fossiliferous sites are to be found just across the Hudson River from New York City, in part at the base of the Palisades and in part just behind the Palisades where the volcanic rocks and the overlying sandstones and shales slope to the west to the Hackensack Meadows.

During the war years I had poked about in some of these New Jersey exposures, particularly in a huge, abandoned excavation known as the Granton Quarry, just a few miles downriver from my home and only about three miles, in an air line, to the west of the American Museum. There I found the bones of small Triassic reptiles intermingled with the abundant remains of fresh-water fishes. Evidently this was an old lake, in and above the quiet waters of which many animals had lived and died. (The exploration of these rocks has continued since the war, and some very interesting fossils have been found. Mention might be made of a little lizardlike reptile with greatly elongated ribs, evidently to support a membrane of skin forming a wing, thus making it possible for the animal to have glided from tree to tree within the ancient Triassic forests. This small aerial reptile, which I described and named *Icarosaurus*—after Icarus of Greek legend who escaped from Crete on man-made wings, only to fall into the sea and perish—and a related genus of the same age from the Bristol Channel region of southern England represent the first attempt at aerial locomotion among the vertebrated animals. It has its modern counterpart in *Draco,* the "flying lizard" of the Orient.)

Work in the eastern Triassic, although yielding very important materials, can be very frustrating, especially because so much of the surface rock is covered with buildings and vegetation, so my eyes were turned to the Southwest, where rock exposures are of enormous extent. In the first summer after the war my attention was focused on eastern Arizona, on the region around Saint Johns, and particularly on the Petrified Forest, where I had a government permit to search for fossils. In making this first postwar

The cliffs at Ghost Ranch, New Mexico . . . rise in colorful splendor—yellow and white and red—above the eroded, variegated Chinle badlands at their base.

sally into distant fossiliferous fields I had to abandon temporarily my family in New Jersey. Our fifth son, Charles, had been born in February, and although Margaret and I had had some thoughts of taking all of the boys to the Southwest for the summer, the idea did not seem very practical when the time for departure rolled around. I went to Arizona, where with Bill Fish and George Whitaker, two of the technicians in our paleontological laboratory, a foray was made in the Triassic Chinle Formation. Some good fossils were found and collected, which encouraged us to make plans for the following summer.

That year, the summer of 1947, I went west with George Simpson, and for a short time we were together in the field. Then he went his way, which was to search for fossil mammals, while I went mine in continued pursuit of Triassic fossils. I was accompanied by George Whitaker and by Thomas Ierardi, a New Jersey neighbor and close friend who wished to participate in the fossil hunt.

The cliffs at Ghost Ranch, New Mexico, are of stupendous beauty. They rise in colorful splendor—yellow and white and red—above the eroded variegated Chinle badlands at their base, and they extend for miles along the eastern side of the Chama River valley. Above the sheer cliffs, formed of the Entrada and Todilto formations, which are of Jurassic age, are the forested slopes of the Jurassic Morrison and the Cretaceous Dakota formations. Thus in this area the Mesozoic sequence is represented by Triassic, Jurassic, and Cretaceous rocks, making the region of particular interest to the geologist and paleontologist. I knew that Triassic fossils were to be found here in the Chinle badlands because the University of California had worked in this region before the war. As we approached Ghost Ranch we were more than impressed by the color and by the dreamlike atmosphere of the scene. The great resplendent cliffs became ever greater as we drew closer to them, until at last we were at their very bases—and at Ghost Ranch. Here was a group of low adobe buildings, sheltered from the June sun by large cottonwood trees and separated from the dry Chinle hills by a large field of alfalfa. It was truly a verdant oasis in the desert.

As we climbed out of our jeep a wiry, sandy-haired man approached us with a pleasant smile. He greeted us cordially and identified himself: Arthur Pack, the owner of Ghost Ranch. It was lunchtime and we were invited in for an excellent spread with Arthur's wife, Phoebe, and the Pack children, Charles and little Phoebe. Also present were Herman Hall, the ranch foreman, and his wife, Jimmy, who had cooked the meal. There we made plans.

Arthur was pleased to have us explore across his thousands of acres of ranchland, for he was especially interested in the fossils that previously had been found here. We made camp in a grove of cottonwoods not far from the ranch buildings, and our fossil hunt began. It didn't last long; within hours we had found a phytosaur skull, and within days we made a discovery that was to change all of our plans for the summer.

It happened on the fourth day of our sojourn at Ghost Ranch. On the morning of that day we separated and went our individual ways, the better to search the gullies and talus slopes of the Chinle beds at the bases of the cliffs. All morning we climbed up and down and back and forth looking for fossils, and for me, as well as for Tom Ierardi, it was not very rewarding work. We found scraps of fossil bones but nothing of significance. Then just before lunch we joined George Whitaker, who had some exciting news. He had discovered a veritable bone bed, bones exposed in a band of outcropping rock halfway up a talus slope, with numerous eroded fragments sliding down the soft embankment. George had some samples to show us.

They were an electrifying sight, those few bone fragments in George's outstretched hand, for they were the bones of a little dinosaur, instantly recognizable from their structure and form. Of special significance was a little piece of a claw, which because of its small size and compressed shape could belong to nothing else but *Coelophysis,* a very early dinosaur that had been discovered in this region almost three-quarters of a century ago. George assured us that there were riches galore in the bone bed he had discovered, so we immediately accompanied him for a look at the place, after which we went to lunch in a very excited frame of mind.

Coelophysis had been described in 1887 by Edward Drinker Cope, the pioneer American paleontologist, upon the basis of a few fragmentary bones found in the Chinle Formation near Cerro Blanco, some miles to the west of Ghost Ranch. Since his first description no appreciable additional remains had been unearthed. Now, it would appear, we had a rich deposit to exploit, a deposit that might yield a complete skeleton. It was a stirring prospect, to say the least.

After lunch we went back to the site and began to probe. The more we probed the more we found. We very carefully removed some overburden back a couple of feet into the hill and along the exposed edge of the outcropping for several feet. Then we carefully worked down to what we hoped would be a layer of bones. It was. Bone after bone came into view—leg bones and vertebrae and ribs. It was significant that these bones were articulated. Soon we began to get indications of skulls. Here was no hodgepodge of isolated bones, but rather the remains of ancient reptiles with all of their bones in place. This was something to think about!

Our first thoughts had to do with plans for the future. Ghost Ranch had been intended as a stopover for a week or two on our way to Petrified Forest, for another season of work there. Here was a discovery that would cancel all of those plans. The paleontologist has to work according to plan, yet he has to be an opportunist as well; he never knows, until they are revealed, what significant fossils may suddenly come to light. When they do appear, he has to collect them, he has to take advantage of the opportunity at hand.

I had to write to the people at Petrified Forest to inform them that I would not appear during the summer, as planned. This was understandably disconcerting to them, because the necessary permits had been obtained, government wheels were turning, and I was expected. All I could do was to apologize and inform them of the facts.

The discovery also changed our life at Ghost Ranch. When we told Arthur Pack about it he promptly decided that we could not spend the summer in a tent. He graciously offered us the use of a house on the ranch—a big house with a large, cool living room, a dining room and kitchen, and upstairs two bedrooms, each with its bath. We were quickly installed in these luxurious quarters, to spend a summer that was never to be forgotten by any of us. It should be added that a large swimming pool was a part of our perquisites. It should further be added that all of this was within a half-mile of our fossil site.

It was delightfully pleasant in the houses and under the trees at Ghost Ranch, but it was very hot indeed on that slope beneath the cliffs where we were to dig for fossils. For the first week or so we rigged a tarpaulin to protect us from the blazing southwestern sun, but soon it became apparent that this could be only a temporary measure. The tarpaulin needed con-

Preliminary work at the Ghost Ranch Quarry. We had a lot of digging to do before we could really start the serious work of getting the fossils out. The big rock above us, which we eyed apprehensively from time to time, never did come down.

stant adjusting and in the afternoon, when the winds came up, it would flap and billow, constantly threatening to come down on top of us and our precious fossils or to sail away across the juniper-covered ravine below our quarry. Consequently, we soon went into town, the town being Espanola, purchased the necessary lumber, and constructed a solid roof over our heads. That made our work much more pleasant.

Also we soon realized that we needed more help if we were going to get some of the fossils out before the end of the summer, so we telephoned New York and asked for Carl Sorensen, one of the senior members of our laboratory force and an old hand at fossil digging, to come out and join us. Carl came, and then there were four of us to dig at the Ghost Ranch Quarry. Even with our augmented force we had a colossal task ahead of us.

Now that the bone layer had been established, and we were convinced that it contained articulated skeletons, the first step was to enlarge the quarry. The fossiliferous layer extended back into the hill, not quite horizontally but with a slight dip downward—perhaps five degrees—as we followed it back. At the same time, the steep hill increased the depth of excavation rapidly as we worked our way back along the bone layer. The result was that we had a lot of digging to do before we could really start the serious work of getting the fossils out. So for days we dug, and we dug. Finally, however, we had our quarry established to within a few inches of the bone bed, going back perhaps fifteen feet or so into the hill, along the *dip* as it is known geologically, extending some thirty feet or more along the face of the slope, along the *strike* of the sediments at right angles to the dip.

From now on the work had to be careful and exacting, and as work went forward the extent and the richness of our fossil find became ever more apparent. What we were bringing to light was an almost unbelievable deposit of articulated dinosaur skeletons, one on top of another, crisscrossed in a most intricate fashion. It developed into an impressive sight.

We were not the only people to be impressed. Arthur and Phoebe Pack and their children and the other folks at the ranch were our frequent callers. People who were visiting the ranch also came up to see what we were doing. Another interested visitor was Georgia O'Keeffe, the eminent artist, who lived not far away. As is well known, she has long been fascinated by the beauty of bones—of skulls and other bones of animals that have dried and been bleached by the desert sun. She was also fascinated by the very different bones of our little dinosaurs.

Coelophysis was a little dinosaur, no more than eight feet or so in length.

Carl Sorensen undercutting one of the big blocks at Ghost Ranch. The work had to be careful and exacting.

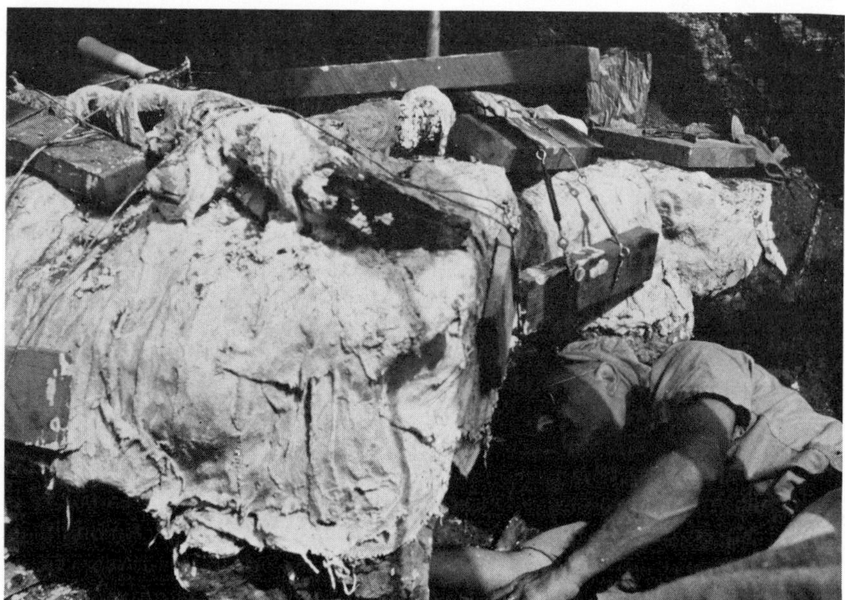

It was a lightly built reptile with a long neck, a lithe, supple body, and a long, very flexible tail. The hind limbs were elongated and birdlike, and served in life to carry this little dinosaur across the ground at a good clip, probably much as an ostrich drifts across a modern landscape on its long, strong legs. The front limbs of *Coelophysis* were short and in no way were used for locomotion. The hands had three clawed fingers, which probably assisted in the capture of prey. That *Coelophysis* was carnivorous is amply indicated by the elongated skull and jaws, furnished with numerous sharp teeth. This little dinosaur, probably weighing no more than seventy-five or a hundred pounds at the most, and living at the very advent of dinosaurian history, must have been a predator upon small game—other little reptiles, perhaps even insects.

Within the body cavity of at least one of the specimens excavated at Ghost Ranch are the bones of a young *Coelophysis*. Is it possible that this dinosaur gave live birth to its young, as do some modern reptiles, rather than laying eggs? Probably not, because the opening in the pelvis is much too small to have afforded passage for an embryo as large as indicated by the bones in the body cavity. It seems much more likely an opening for the passage of an egg. Moreover, the bones are those of more than an embryo, rather of a young dinosaur that had been for some time living its own life in the great world. The inescapable conclusion is that *Coelophysis* was cannibalistic, eating its own young on occasion, just as do some modern reptiles. Not a pretty picture, but a realistic one.

How did so many skeletons happen to be concentrated in this spot? We can only speculate as to this. Perhaps the animals, traveling in a considerable herd, were trapped in some sort of a quagmire or quicksand and perished. The fact that the skeletons are so nicely articulated would argue against their being washed downstream for any significant distance. Being small, light animals, they may have drifted for a short way, to be caught and dropped in an eddy or backwater. Certainly they were not rolled over and over in bone-breaking and bone-separating turbulence. They must lie about where they died. Whatever the cause, we were rewarded by an unusual number of dinosaur skeletons—not large dinosaurs, but certainly among the most completely preserved dinosaurs ever found.

Although, as previously remarked, quarry work is not necessarily exciting, nevertheless the thrills are there—the inner joys of seeing new specimens appear under the ministrations of pick and awl and brush. There were also other satisfactions in our work: the pleasure of watching the light from the wheeling sun change and accentuate in different ways the colors of the cliffs that rose above us just across a little ravine from where we were working; or the pleasure of watching the circling of hawks and ravens in the blue sky overhead; or the pleasure of seeing a deer or perhaps a pronghorn antelope on the hilly slopes below us.

George Whitaker on the sled. We hoisted the blocks on the sled and then Herman pulled it with the bulldozer to the ranch houses. It took quite a few trips to complete the job.

The summer days passed by, and in time we had a large quarry floor, covered with completely mixed skeletons. Now we were faced with some difficult decisions. How were we to cut channels through this great platform of skeletons in order to isolate blocks of such size that they could be removed? We agonized over this for many hours, realizing that in the process of cutting the channels we inevitably would have to cut through some bones. There was no way around it.

We began our work, and as we cut down with small picks, hammers, and chisels (carefully removing as many bones as possible in the paths of the channels) we found that there were skeletons not only on the quarry floor that we had exposed, but also skeletons underneath, layer after layer. We would start a channel at a promising-looking place, a place seemingly devoid of bones, only to find at a little depth a valuable skull or pelvis right in the path of the channel. Such a bone would have to be extracted, or in some manner bypassed. All of this made the channel cutting the most difficult and time-consuming of our quarry operations. It goes without saying that during the quarrying there was a constant keeping of records—of photographs and maps and notes to show the disposition of the skeletons and of individual bones within the quarry. In time we had our channels

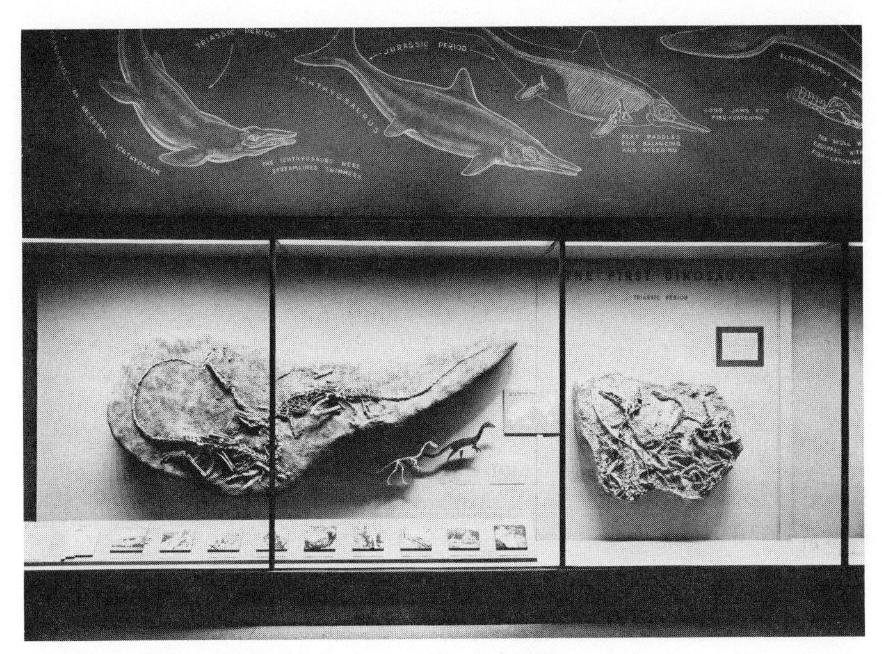

Some of the skeletons of Coelophysis *are to be seen at the American Museum in New York. Above the exhibits is part of a chalk mural; these murals are described in Chapter 9.*

cut, so that the quarry floor, once inviolate, was divided into a series of large blocks ready to be plastered and turned in the usual manner.

How were such heavy blocks to be moved? This was where Arthur Pack came to our rescue. He had a large Caterpillar bulldozer, and Herman Hall loved to operate this machine. Herman bulldozed a road to our quarry, the last few yards of the trackway being so steep that the bulldozer was almost standing on end as it reached us. While Herman was doing this we had devised a heavy wooden sled. We hoisted the blocks on the sled, and Herman with the bulldozer then pulled the sled to the ranch houses. It took quite a few trips to complete the job, but finally we had a large pile of blocks ready for shipment to New York.

Shipment was to be by truck. A heavy vehicle was leased and the blocks were winched on to it for their final journey. So ended the summer at Ghost Ranch.

Epilogue

There was another summer of collecting at Ghost Ranch by George Whitaker and Carl Sorensen. I could not be present; that year I was attending the International Geological Congress in Britain.

As a result of the two summers of collecting a considerable number of fine skeletons of *Coelophysis* were obtained for study and exhibition. Some of them are to be seen at the American Museum in New York; others are visible elsewhere because we supplied various museums with blocks that we had collected. One block is on display near its point of origin, at the Ghost Ranch Museum.

A few years later Arthur Pack, a public-spirited man of unusual foresight, presented Ghost Ranch to the Presbyterian Church, to be used as a conference center. Under the able direction of the Reverend James Hall it has become an institution of international repute, where people gather for conferences and seminars on a variety of topics, ranging from theology to sociology to science to art and beyond. Ruth Hall, Jim's wife, has long been interested in the Ghost Ranch fossils and has collected some fine specimens for preservation at the ranch.

Georgia O'Keeffe still lives in her house near the ranch. She is now (as of 1979) in her nineties, and still a person of great vitality. Several years ago Arthur died; his body reposes in a grave on a little hill behind some of the original Ghost Ranch buildings. There he rests in peace, at the place that he loved so dearly, and where he spent some very happy years.

In 1977 the Department of the Interior designated the Ghost Ranch Quarry as a Natural Landmark. One fine afternoon a ceremony was held under the trees at the ranch, attended by a goodly crowd of people. A representative of the government was there, to present a bronze plaque, signifying the importance of Ghost Ranch. Speeches were made, refreshments were served, and the whole crowd hiked up to the old quarry site, now completely filled in by slumping, to see the place where the little dinosaurs of Ghost Ranch had been found. Tom Ierardi was present and I was, too, along with Margaret. George Whitaker was unable to be present. Carl Sorensen had died several years previously.

9. EXHIBITING, EDUCATING, EDITING

This is to be a chapter about things other than research and fieldwork, and of these things exhibition comes first. Exhibits were what first aroused my interest in paleontology, and in museums. Back in those boyhood days at home I had my little museum—the glass case filled with an eclectic array of fossils, shells, minerals, arrowheads, coins, and various other objects that had seemed precious to me. In the summer after my graduation from high school I had visited the Field Museum in Chicago, where I was awed and intrigued by the displays to be seen in that vast building. In following summers, on my way to and from Colorado I had visited the museum at the University of Nebraska, where the elegant skulls and bones of Cenozoic mammals fired my imagination. To me paleontology was in part, and it was a large part, a matter of putting fascinating objects out to be seen. It still is.

During my earlier years at the American Museum I was involved with exhibits in a small way. Then came my first big assignment, in 1953, when it was decided by the museum administration that one of the two dinosaur halls should be revamped. Such a revision was long overdue, because as a result of depression years followed by war years, nothing of consequence had been done to update these two galleries.

A good museum should always be in a state of flux, a good exhibition hall should not long remain static. Desirable revisions of the exhibits are necessary if a museum is to be a vital institution. The revision or renovation of an exhibition hall requires much effort and usually the expenditure of a considerable amount of money, and this is one of the problems of museum practice. Some types of exhibits are much easier to change than others. Generally speaking, the installation or revision of an art gallery is comparatively simple; it is a matter of fresh paint, perhaps revised lighting, and the judicious arrangement of pictures on the wall or sculptures on the floor. A natural history or historical room is something else again. There may be large, sometimes immense, objects to be exhibited, and generally there are cases to be filled—tasks that more often than not require an unbelievable amount of work. There are labels to be written. The prospect of doing over a large hall is thus a daunting one to say the least.

As I contemplated the hall that was to be done over—called the Brontosaur Hall—I was indeed daunted to some degree. I was challenged as well. It was a big hall—no doubt of that—some 150 feet in length, 60 feet wide, and 24 feet from floor to ceiling. It was all open space; there were no pillars. Along the sides and across the ends of the room were built-in cases consisting of plate-glass fronts with bronze fittings. In the middle of the room were dinosaur and other reptilian skeletons, dominated by a massive brontosaur, 60 feet or more in length and 16 feet high. To cope with this problem I was to work with some of our paleontological laboratory force, as well as with the exhibition department of the museum, then under the direction of Katherine Beneker.

Working with Kay Beneker was indeed a pleasure. She was a person of exquisite taste and good judgment. We embarked upon our task with Kay deciding on the aesthetic impact of the hall, while I had charge of its paleontological contents, which was and is the usual arrangement at the museum. The curators are responsible for the systematic and intellectual arrangement of the exhibits, while the exhibition department is responsible for the presentation of the displays.

We decided to group on a common base three of the big dinosaur skeletons, heretofore lined up separately like freight cars in a train shed. Moving a 60-foot brontosaur is not easy, but fortunately all of the skeletons at the museum had been mounted on bases supplied with casters, so it was a matter of pushing and shoving with electric trucks supplemented by manpower until we had the skeletons arranged the way we wanted them in the middle of the hall. In addition to *Brontosaurus,* this group included the plated dinosaur, *Stegosaurus,* and the giant carnivore, *Allosaurus.* The idea was to build a free-form base around these huge skeletons, and that was done. But more than this was involved.

Immediately before the war Roland T. Bird, who was then on the museum staff, had made an unusual and spectacular collection of gigantic dinosaur footprints along the Paluxy River, near the little town of Glen Rose in central Texas. These were the trackways of a huge brontosaur that evidently had been pursued by a large carnivorous dinosaur. R. T. (as he was affectionately known to all of us) had done a superb job of collecting these footprints. He had taken out, piece by piece, a segment of trackway about 16 feet in length, preserving twelve brontosaur footprints, six of the front

We decided to group on a common base three of the big dinosaur skeletons. Brontosaurus *in the left foreground,* Stegosaurus *to the right, and* Allosaurus *in the rear.*

feet and six of the hind feet, as well as six carnivore tracks, all made by the hind feet (since the carnivorous dinosaurs were bipedal).

The tracks made by the brontosaur were as large and as deep as big washtubs; the tracks of the carnivore were proportionately large. The tracks were preserved in a marly limestone, what had once been the muddy floor of a shallow lagoon near the seashore, where these reptilian giants, the hunted and hunter, had splashed through water perhaps four or five feet in depth. These trackways were particularly exciting, because they recorded a drama that took place perhaps 80 million years ago. It is clear that the carnivore was following the brontosaurian giant; the tracks of the hunter parallel those of his presumed prey, and at the front of the trackway one can see where the carnivore stepped into the footprints of the brontosaur. All of this is preserved in stone as clearly as if it had taken place yesterday, even to the ridges pushed up around the brontosaur footprints, caused by the tremendous weight of the animal as it stepped down in the soft mud.

We had the trackways in storage in the form of more than a hundred big blocks of stone. How were they to be fitted together and who was to fit them? There was only one answer; the man for the job was R. T., but R. T. had retired to Florida. Could he be lured back for this assignment? He could and was. Indeed, R. T. was delighted to come and put the tracks in place; it was the fulfillment of a long-held dream of his.

Would the floor maintain the immense weight of this stony trackway? We sought the advice of architects and engineers. The floor would hold up, they said. Where were we to put the trackway? Obviously the place for it was within the big freeform base, immediately behind the brontosaur skele-

Preserved in marly limestone perhaps 80 million years ago, these two sets of prints may have recorded the dramatic moment when a carnivorous predator (left footprints) caught its prey, a brontosaur.

ton, but there was a problem; the long tail of the brontosaur was in the way. So our laboratory technicians dismantled the tail and remounted it curving to one side, as if the big dinosaur had swept his tail in that direction as he walked along. Space was thereby provided for R. T. to assemble the trackway, which he did with consummate foresight and skill. It took him several months, but in the end the job was successfully finished, resulting in a most spectacular addition to the center of the hall.

In the meantime we obtained the services of a lighting expert and he determined where floodlights should be placed in the ceiling, the most effectively to illuminate the big central island with its dinosaurs.

While all of this was going on, we were busy installing displays within the cases that lined the periphery of the hall. In between times I was busy writing labels, which is not as easy as may be supposed. Good labels must be informative yet succinct, and that requires much writing and rewriting.

Above the side and end cases was a staggering amount of wall space—about 10 feet in height and the cumulative length and width of two sides and two ends of the hall. There had long been thoughts of murals depicting the age of dinosaurs, but considerably more than 400 running feet of wall space, 10 feet high, calls for a lot of painting. Could it ever be done in our lifetimes? Then in conversation with Bert Parr, the director of the museum, an idea was born, an idea of remarkable practicality and simplicity. Why not paint the walls a deep blue as a background, and then fill them with outline sketches in white, like drawings on a blackboard? That was what we did.

I worked with some of the museum artists and we produced sketches to scale of the subjects that were to occupy the walls, sketches of dinosaurs and other reptiles that lived in the Mesozoic past. These were made into lantern slides and the images were projected on the walls, which in the interval had been given their deep blue coat of paint. Then it was merely a matter of tracing the projected images on the walls. The tracings were done with white chalk, and we intended to paint over the chalk lines with white paint. But when the chalk images were completed they were so very elegant (the chalk lines had a soft texture that could never be reproduced in paint) that we decided to leave them as they were. Should we fix them? No, we decided, fixing might dull the chalk lines. And as the director pointed out, the chalk drawings were up where they would not be disturbed, and under such conditions they were reasonably permanent. So we left them, and a new technique, that of chalk murals, was born.

There is no reason to go into further details about our work on that hall. There were a lot of details, and they took a lot of time. Also there were the usual sets of mishaps that go along with such a project. Such as the time a painter, applying his brush to the ceiling, walked off the end of a scaffold and plunged down through the top of one of the wall cases, to land on the skeleton of a small dinosaur therein. Miraculously he was not seriously hurt—a few cuts and bruises—but the dinosaur suffered in a large way (in time it was repaired). Or the time a large dinosaur leg bone got away from a couple of the lab men and crashed through the plate-glass front of another wall case. Still another repair job on bone and glass.

At last the hall was completed and there was a grand opening. It had taken the better part of a year of my time, but it was an interesting project and a grand experience. In the years since it has been a great pleasure to see museum visitors of all kinds, large and small, young and old, casual and serious, enjoying the hall.

Several years later we tackled the other big dinosaur hall, known as the Tyrannosaur Hall, containing the impressive array of skeletons and skulls that had been collected largely by Barnum Brown through the years. Again, it was a project that involved Kay Beneker, men from our paleontological laboratory, the exhibition department, and myself. But this time the problems were somewhat different than those we had faced when we began work on the Brontosaur Hall.

The Tyrannosaur Hall dated back to the early twenties, when the museum wing in which it is housed had been built. It is a large, open hall, as is the other dinosaur gallery, but instead of internal lighting it is illuminated by high windows on each side of the room. Our first decision therefore had to do with lighting: Should we retain daylight in the hall, or should we block the windows and install internal lighting in the exhibition cases along the sides and ends of the hall? This latter course would have been very expensive, so we decided to stick with daylight and I think it was a good decision. It made a pleasant break for the public, contrasting as it did with the halls at either end of this gallery, which were internally lighted. The visitor would come from darkened rooms with dramatically lighted exhibits into a bright, sunny room in which there was a feeling of the world outside. Moreover, the natural light brought out the colors and textures of the dinosaur bones in a very pleasant manner. My one disappointment in the matter of lighting was that we never got the draperies we had wanted for the windows, to soften the light.

As was the case with the other dinosaur hall, we decided to group several large skeletons in the middle of the room: skeletons of the gigantic carnivorous dinosaur *Tyrannosaurus,* of the horned dinosaur *Triceratops,* and of two duckbilled dinosaurs *Anatosaurus.* One fine morning we all assembled in the hall with a gang of men from the custodial department, to shift the skeletons on their wheeled bases into their newly appointed places. When we arrived we were greeted by the sight of a large class of art students from the Pratt Institute in Brooklyn, there on an assignment. They were seated all around the hall, making charcoal sketches on large drawing pads of various dinosaur skeletons. Naturally, when we got ready to move the first skeleton, which was *Tyrannosaurus,* there was a great outcry from a part of the class; these earnest students were only partway through their sketches of the skeletons, and now we were going to alter their viewpoints right in the middle of the project. There was no help for it; we had a group of husky men there, appointed for the job at this particular time, so we had to go through with it. I did sympathize with the frustrated students.

In this hall we enclosed the skeletons within a rather formal base, covered with specially fired black tiles. It made a most handsome centerpiece, a focal point in the hall, the decor of which was a light green. We did not attempt murals here. Rather, we installed simple titles and descriptions in large white letters on the green walls, applicable to the nearby exhibits. In

Skeletons of the gigantic carnivorous dinosaur Tyrannosaurus, *the horned dinosaur* Triceratops, *and two duckbilled dinosaurs,* Anatosaurus, *made a most handsome centerpiece, a focal point in the hall.*

designing the descriptions we interspersed pictures of dinosaurs and other forms of Mesozoic life within the text. It proved to be very effective.

Once again, the work on this hall occupied the better part of a year, it was a labor of love as well as of necessity, and it had a satisfactory result. In the years since the hall in its present form was completed it has been visited certainly by millions of people, and I trust that most of them have enjoyed their visit. A quarter of a century and more has now passed since we revised the two dinosaur halls, and as this is written they are surely ready for a new revision. I hope it can be done in the not too distant future.

Exhibition is one form of education, a teaching technique that ideally reaches a large and varied public. The exhibits at the American Museum have traditionally been designed for many people, for the casual or not so casual museum visitor, for the school child and for the serious student, whether he or she be in a university class or one of the numerous people now involved in adult education activities. And it was very satisfying to

Dr. Dale Russell now of the Geological Survey of Canada (left) and Dr. John Ostrom now of Yale University (right) were two of my students who went on to become outstanding scholars. It was a pleasure to have these alert students, learning and helping me to learn.

participate in this sort of education. For some twenty-five years after the war I was also involved in formal university teaching, a program that occupied a considerable part of my time.

As I said earlier, Professor Gregory taught at the museum the graduate courses in vertebrate paleontology that were a part of the Columbia University curriculum, and he also supervised the research of doctorate candidates in this field. He had been engaged in this activity since 1910, when Professor Osborn had retired after some twenty years of active teaching. At the end of the war, Dr. Gregory had retired, after having trained an impressive number of professional paleontologists.

Plans were then made for George Simpson and me to take over Gregory's university work—Dr. Simpson to give instruction on the evolution of mammals and birds, I to cover the evolution of the so-called lower vertebrates, the fishes, amphibians, and reptiles. We were both appointed to professorships in the Graduate Faculty. The general plan was for me to present the course on lower vertebrate evolution, to be followed by Simpson on the mammals. We would each personally supervise the work of students pursuing higher degrees in our fields. By agreement with Columbia, the museum furnished a well-appointed lecture room and laboratory as well as facilities for students doing research. It was a case of Mahomet going to the mountain.

We began our program of joint instruction and everything went well, except that I had to initiate the project under a rather intimidating psychological handicap. Simpson and I had agreed that during that first academic year he would sit in on my lectures and I would do the same with his sessions, in order that we could properly integrate our presentations. What should then happen but that Ernst Mayr, the world-famous ornithologist and evolutionist (he was then at the American Museum), and Theodosius Dobzhansky, the equally world-famous geneticist and evolutionist (he was then at Columbia), decided to join Simpson as auditors. There I was facing very bright and knowledgeable graduate students in front and the battery of Simpson, Mayr, and Dobzhansky in the back. The students unhesitatingly disputed me on every possible occasion, and the big guns in the back shook their heads in disagreement whenever they felt that I was going off the deep end. At any rate, I got through the course, and then it was my turn to sit in the back row with Mayr and Dobzhansky. I was comforted to see that Simpson had as hard a time with the students as had I, and that Mayr and Dobzhansky shook their heads in disagreement on more than one occasion.

My years of graduate teaching were for me years of learning, for there is nothing like a group of graduate students to keep one on one's toes and abreast of new developments in the field. It was a pleasure to have these alert students, learning and helping me to learn, and to see many of them go on to become outstanding scholars.

My learning also was immensely advanced and abetted by the many years of association with my close paleontological colleagues at the museum, notably Bobb Schaeffer, the Curator of Fossil Fishes; Malcolm McKenna, who took over fossil mammals when Dr. Simpson left the museum for Harvard; Richard Tedford, who came a few years before I retired from the museum to work on fossil mammals; and Norman Newell, an invertebrate paleontologist and a long-time friend. Many were the discussions that took place between us on problems of animal evolution and distribution. Many were the experiences that we shared.

As we were finishing the Brontosaur Hall I thought it would be a good idea to publish an article somewhere, to present the hall to the museum profession, yet it seemed that there was no suitable vehicle in this country for such an article. In the end I sent a little manuscript to the *Museums Journal* in England, where it was published through the kind offices of my old friend Bill Swinton. That got me to thinking: Why should there not be a good journal for museum professionals in this country? I talked the matter over with the director of the museum, and we both agreed that the lack of such a publication here was truly a deficiency that should be corrected. The American Association of Museums did publish at that time *Museum News,* but it was a small information sheet more than anything else.

Consequently, Parr and some others at the museum, as well as myself, decided to try to get something going in the Museums Association. It seemed, however, that the powers that be in the association at that time were against any such effort. Perhaps they did not wish to be bothered. Certainly, when I rose at an annual meeting with such a proposal, the idea was not received up front with enthusiasm. However, a committee of three was appointed, of which I was one member, to look into the matter and report back.

The next thing I knew, some months later, the other two members of the committee had framed a report to the effect that the creation of a new publication for the Association of Museums was not feasible. When I saw their report I insisted on appending a dissenting opinion, which I did. The decision was two to one against me. I therefore suggested that if the American Association of Museums did not wish to establish a publication for museum affairs the field was open to some other organization, and the other committee members agreed to this. *Curator* was founded, to be published by the American Museum of Natural History.

Bert Parr was very much for such a publication. He was joined by a group of us at the museum, notably Lester Aronson, Bill Burns, Ruth Norton, Gordon Reekie, John Saunders, Harry Shapiro, Lothar Witteborg, and myself. We had a series of planning sessions, during which the name *Curator* was chosen for the journal, and steps were taken to assure financing by the museum. Then the first issue, of which I was the editor, ap-

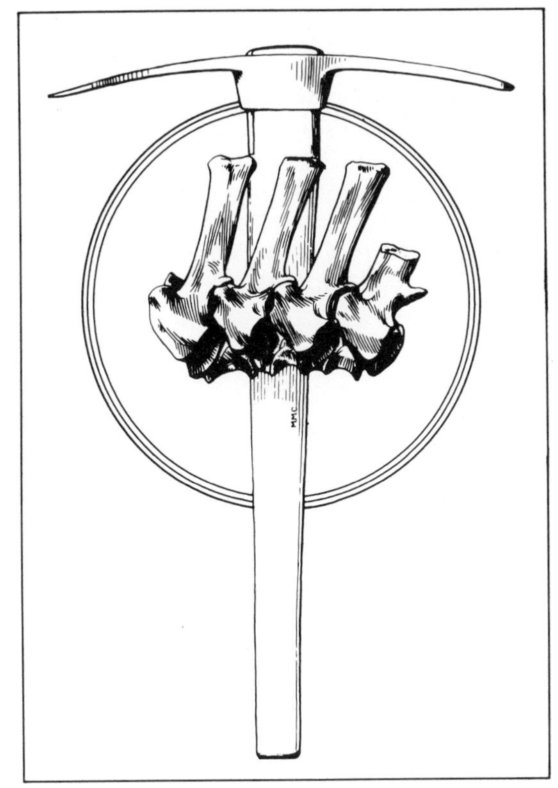

Margaret came up with a most handsome design for the cover of the Bulletin, *and it has appeared on every issue since then. The vertebrae are of the Permian amphibian,* Eryops. *Behind them is a Marsh pick, the universal field tool of vertebrate paleontologists. The circle ties it all together; perhaps it may be regarded as the band of V. P. brotherhood.*

peared in 1958. It was a very handsome journal, thanks to the artistic supervision by Gordon Reekie, and it has remained so to the present day, for be it known that *Curator* is still serving the museum profession and serving it well.

I functioned as editor for the first five years of its life, and then I asked to be relieved, for I felt that some other person should continue with the work. Harry Shapiro, chairman of the department of anthropology at the museum, took over for the next five years; since then *Curator* has appeared under the supervision of Thomas Nicholson, presently director of the museum.

Quite a few years previous to the beginning of *Curator* I had a hand in the development of another publication—the *News Bulletin* of the Society of Vertebrate Paleontology. The society had been founded on December 28, 1940, and at the beginning a mimeographed bulletin was published, thanks to the efforts of George Simpson. Then he went off to war.

I seemed to have inherited the *News Bulletin,* and I decided to try to put it into a more professional type of format. Dr. Brooks Ellis was then at the museum in charge of a large project, a worldwide Catalogue of the Foraminifera (these being microscopic organisms, fossil and recent). As a part of this project he had printing facilities at hand. I got together with Brooks, who was a close friend, and between us we devised a format for the *Bulletin* that could be printed by offset. We made plans for the first issue in this new garb—a 6- by 9-inch publication—and I dragooned Margaret into designing a cover.

This she did, working one autumn Sunday afternoon in the sun on our back terrace at home, suffering from a vicious cold, and with our small children demanding attention all of the time. She came up with a most handsome design for the cover, and it has appeared on every *Bulletin* since then. (The logo she designed for the *Bulletin* cover was eventually adopted as the official emblem of the society, and a very popular emblem it has been.) The format of the *Bulletin* and the manner of presentation of its contained materials has continued to the present day. The publication is still going strong after more than thirty-five years of service to the profession of vertebrate paleontology, worldwide.

One other editorial stint might be mentioned. For several years I edited the journal *Evolution,* the publication of the Society for the Study of Evolution, an international scientific group which I served as president for one year.

10. THIS OTHER EDEN

It was a bitterly cold morning on January 3, 1959, when Margaret and I boarded a Varig plane at Idlewild Airport (as it was then) to fly from New York to Rio de Janeiro. At last, after having collected and studied Triassic amphibians and reptiles in North America, I was to have the opportunity of crossing the Line, to have a go at some of the ancient animals of South America. I was to be a guest of the Conselho Nacional de Pesquisas of Brazil, a body corresponding more or less to the National Science Foundation of the United States, and I was to work with some Brazilian friends—notably Carlos de Paula Couto, Fausto Luiz de Souza Cunha, and particularly Llewellyn Price, a bonafide Brazilian in spite of his Welsh name. (Llew had been born in Brazil of American parents, so he had dual citizenship. After an education in the United States he opted for Brazilian status.) Ours was a journey long desired and long anticipated.

The flight down was memorable. Those were still the days of propeller planes, and we were on a Constellation that took a good many hours to make the trip, with some landings along the way. We were treated to luxurious accommodations and marvelous food; airlines don't do that anymore. As we came into Rio, Margaret was moved to tears by the beauty of the great bay and the almost unearthly rugged hills, or morros, that rose verdantly in and around the city. It was the beginning of our love affair with Brazil—that other Eden of the tropics.

Llew and Carlos met us at the Galeão airport, and installed us in a comfortable little hotel facing the ocean at Copacabana, where fashionable Carioca ladies strolled along the sand in the briefest of bathing suits, where the men and boys played endless games of soccer across the beach, where lovely kites shaped like birds tugged at their strings in the tropical sky above the strand, and where the great surf of the South Atlantic pounded in with gigantic percussions that made themselves felt even in our room. Could we ever forget the breakfasts on the little balcony overlooking the beach and the ocean?

A week or so was spent in Rio getting acquainted and making various preparations before we were to leave for Rio Grande do Sul, the southernmost state of that immense country. It was then that we renewed our friendships with the Price and Paula Couto families (they had previously been in New York) and got to know other Brazilians who became our fast friends. Rio de Janeiro has been described countless times; there is no need to go into ecstasies at this place. Mention should be made, however, of our days at the Museu Nacional, which was the former palace of Dom Pedro, once the Emperor of Brazil, a still ornate building set in a large park, the place where Carlos worked, and at the Divisão de Geologia e Mineralogia on Botafogo Bay, where Llew held forth. These were locales familiar to the young Charles Darwin when he was in Brazil in 1832 during the voyage of the *Beagle,* though in Darwin's day the scene was semi-rural, with charming baroque-style houses and small buildings where now great skyscrapers rise to vie in height with the nearby morros.

Thanks to Carlos we had a nice trip around the city and into the surrounding country, including a leisurely stop at Alta da Boa Vista, where a small river slid down over the rocks, where tree ferns of Mesozoic aspect stood against the sky, and where great tropical butterflies followed erratic flight patterns through the lush forest. One could appreciate Darwin's statement that the most inspiring sight he had ever seen was the tropical forest of Brazil.

All too soon, it would seem, we (Price, Couto, Fausto and ourselves) were ready to leave the delights of Rio for the pampas of southern Brazil.

I don't know what I expected in Pôrto Alegre, the capital city of the state of Rio Grande do Sul and the terminus of our flight south from Rio. Whatever it may have been, I was surprised beyond all measure when we flew into that southern metropolis. Here was no sleepy provincial town, but rather a large and very bustling city, its many tall buildings rising above the Rio Jacuí where this broad river flows into the head of the Lagoa dos Patos—the Lake of Ducks—that extends south for 150 miles from the city to its connection with the South Atlantic. Here we tarried again for several days before embarking on our journey into the interior of Rio Grande do Sul, a journey that had as its destination the city of Santa Maria.

Santa Maria had for us a particular significance because it is the type locality for the Triassic Santa Maria Formation—a band of sedimentary rocks stretching from east to west in a great arc across the northern part of Rio Grande do Sul, then curving down along the western border of that state to enter Uruguay to the south. The Santa Maria beds, more often than not appearing as brilliant red exposures among the subtropical greenery, have over the years yielded many interesting remains of fossil reptiles. These fossils are especially important because of their relationships to fossils found in other countries and southern continents, particularly to Triassic reptiles occuring in Argentina, in southern South Africa, and to the north across the equator, in central India. Here we were to search for specimens additional to those that already had been found as part of the long, cooperative task of assembling the pieces that make up a gigantic picture puzzle: the picture of animal distributions at a time, millions of years ago, when continents were more closely conjoined than they are today and located in latitudes different from their present positions.

During our stay in Pôrto Alegre we spent much of our time at the University of Rio Grande do Sul, where there is a very active School of Geology, and where I gave a lecture (in English) with appropriate gestures and pictures on the blackboard, which I hope made it tolerably comprehensible to the students, all of whom were trying to improve their understanding of English. And through the kind offices of Professor Irajá Damiani Pinto, the head of the geology program, we were provided with a van for our drive to Santa Maria.

It was an all-day drive of about 200 miles over red dirt roads across rolling pampas country, with along the way occasional fazendas, which are large farms, their houses and outbuildings capped with red tile roofs that glowed bright in the southern sunlight. Also we encountered gaúchos—Brazilian cowboys—herding cattle as we drove along. Never will I forget our crossing of the Jacuí River on a primitive ferry. It was an idyllic sight, the blue river bordered by a thick riparian forest, and overhead a bright sky flecked with clouds. Standing in front of us on the ferry was a wandering gaúcho, a picturesque figure in a black, broad-brimmed hat, a flowing pink shirt, very baggy black trousers with the inevitable faca, or knife, thrust in his belt, and bare feet partially encased in heelless slippers. In this last detail he was less than complete, perhaps because he was off duty. When mounted, the gaúchos wear distinctive boots, the tops of which are accordion-pleated, to give them great flexibility.

By late afternoon we arrived in Santa Maria, a small city, yet nonetheless containing in its center a considerable cluster of tall buildings. Right in the middle of town was the praça, the equivalent of the Spanish plaza, and this little park with a bandstand in its center was the focus of evening activities. Here the citizens would gather to hear the band play, and the little boys and girls, dressed in their finery, would solemnly perform round games, the girls generally holding the hands of their small brothers and guiding them properly through the simple intricacies of the innocent sport. It was a delightful sight.

We established ourselves in a small hotel just off the praça, and for the

On the ferry was a wandering gaúcho, a picturesque figure in a black, broad-brimmed hat, a flowing pink shirt, very baggy black trousers with the inevitable faca, *or knife, thrust in his belt, and bare feet partially encased in heelless slippers.*

first and only time in my career I did fieldwork via a taxicab. The field car in which we had journeyed from Pôrto Alegre was of necessity assigned to other work, and there we were for the moment without any transportation. Almost every morning we would get a cab in front of the hotel and have it drive us to Quilômetro Tres (there is no *k* in the Brazilian version of the Portuguese alphabet), where we would disembark to spend the day hunting for fossils. (Sometimes, if we were lucky, we would catch a bus.) As may be surmised from its designation, Quilômetro Tres was not very far from the middle of town; indeed, it was just on the edge of things, in what might be called a rural suburb, an area of small landholdings and truck gardens. Here, in miniature badlands where the red sandstones and siltstones were eroded and exposed, we searched for bones. At the end of the day we would get back to town as best we could, sometimes by catching a bus, sometimes by cab if we could get one. It was a strange way to go fossil hunting, but it was sufficiently effective.

It was effective in part because of the abundance of Triassic reptilian bones that were weathering out of the sediments. In some places, what looked like accumulations of pebbles and rocks washed down into the bottoms of little draws (*sangas* in Brazilian) were actually fossil bones and bone

Llewellyn Price and Fausto Luis de Souza Cunha applying bandages to a fossil. Here, in miniature badlands where the red sandstones and siltstones were eroded and exposed, we would search for bones.

fragments. Our task was not so much to find fossils, but to search out good fossils—specimens worth collecting. Locating such fossils was not easy.

For one thing, most of the specimens that had accumulated in the bottoms of the sangas were isolated bones and fragments, and we were looking for associated skeletons or parts of skeletons. Again, the bones in the Santa Maria beds were often preserved in a most unusual manner; mineral-bearing waters had through the ages penetrated the fossils through tiny cracks and openings to deposit minerals which gradually forced the fossil material outward, thereby enlarging the bones and making them appear much heavier and more robust than they actually were in life. We were trying to avoid such distorted bones as much as was feasible, to concentrate on bones that preserved more accurately their original contours and proportions. Many hot, humid hours and days were spent in scratching around in the sangas looking for suitable materials and particularly trying to locate skeletons.

The Brazilians don't have a tradition of carrying lunch baskets along for a noon-day snack, so Llew Price quickly made arrangements for us to have lunch at a nearby farmhouse. It was the farm of Senhor Carlos Hübner and his family, a jolly wife, a son of about twelve years of age, and twin daughters perhaps eight years old. They were good, friendly people, and we had many delightful meals with them. In short order we felt that

Seated (left to right): Fausto Luis de Souza Cunha, Llewellyn Price, the author, and Senhor Carlos Hübner, proprietor of the farm, with one of his twin daughters standing at his side.

The rhynchosaurs were very strange reptiles of Triassic age. (Restoration by Margaret Colbert; from Edwin H. Colbert, The Age of Reptiles, *Weidenfeld and Nicolson, London. W. W. Norton, New York.)*

we were all members of a large, extended family. Our fare was more in the nature of a full meal than a lunch, because Brazilians eat heartily at noontime. We always had feijãos e arroz—beans and rice—and always two or three kinds of meat. So we were well fed.

Senhor Hübner (and incidentally there is no umlaut in Portuguese) was of course a Brazilian of German ancestry, which made him one of a very large minority in southern Brazil. He raised vegetables, which he sold in Santa Maria, taking them into town early every morning in a little two-wheeled cart, pulled by a small horse. Then later in the morning he was back to work his small fields through the remaining hours of daylight.

The remaining hours of daylight were limited, because in that latitude night came on at about six o'clock. Well before that we would be back in town, to have a shower and cool off before we had our evening meal. As the skies became dark and the air grew cooler we would retire to our rooms for a long night, with the sun not to appear over the eastern horizon until about six in the morning. Often I would awaken before it was light and lie in my bed and out the window watch the unfamiliar stars of the southern heavens wheel past.

On the second day after beginning our search for fossils at Quilômetro Tres we located the skull and skeleton of a nice rhynchosaur, such was the abundance of fossils here, and so we began to think about taking it out. The rhynchosaurs were very strange reptiles of Triassic age. Many of them were rather large and heavy—as large as a big pig or even a small ox. They had bizarre skulls, in which the premaxillary bones, which originally had borne teeth, were toothless and enlarged to form huge, rodentlike "tusks." Why living, exposed bone, subject as it is to infection, should be substituted for teeth is one of the puzzles of paleontology. In the sides of the upper and lower jaws were batteries of buttonlike teeth, forming a sort of grinding mill, above and below, that would have been useful for smashing and pulverizing plant material. The body in these peculiar reptiles was very capacious, the legs were heavy and strong—quite obviously adaptations of an animal that ate bulky plant food. It has been suggested that the rhynchosaurs (which are related to the modern little lizardlike tuatara—*Sphenodon*—of New Zealand) may have scratched tubers out of the ground, or perhaps they ate hard-shelled plants of some sort that they could crush with their dental batteries. Perhaps we shall never know just how they lived—they are a paleontological enigma.

The specimen was in an eroded sanga known as Sanga de São José by the side of a dirt road, perhaps a quarter of a mile or so from the Hübner house. Just across the road from our fossil was a row of little square wooden houses, each painted some bright color such as blue or yellow, and each with a peaked red tile roof. The house immediately across from us was the home of Senhor Anton Perrera and his wife. The Perreras, who were middle-aged folks, took quite an interest in our work. Senhor Perrera, an olive-skinned, wiry man with graying hair that was beginning to disappear from his forehead and the top of his pate, seemingly was retired. Clad in loose pajamas and slippers, he sat during much of each day in the shade by his house. Whether retired or not, he certainly was adhering to the old Brazilian maxim for retired gentlemen: "sombre e agua fresca e chinelas largas"—shade, fresh water, and loose sandals. Every now and then he would cross the road to watch us work, and at times to help us. Senhora Perrera was a handsome, quite dignified gray-haired lady. There were usually some grandchildren around the place.

In the middle of each morning and in the afternoon we would retire across the road to the Perrera house for chimarrão, a drink known in Argentina as maté. It is made from the leaves of a tree, steeped in boiling hot water. The leaves are put in a gourd, a *cuia,* the water is poured in on top of the leaves, and then the drink is sipped through a silver tube, a *bomba,* with a strainer on the far end. The *cuia* is passed from person to person while conversation flourishes. It made for us a nice interruption during a hot morning or afternoon.

And there were opportunities every now and then to interrupt our work and relax even in the quarry. Such was the day when an army of leaf-cutting ants appeared, marching in single file, each ant carrying a piece of leaf like a green banner, on its way to the nest. The procession came into our quarry from one side and marched right across the fossil skeleton, to go on its way at the other side of the quarry. We watched them with much interest until the parade had passed by, and then we resumed our digging.

As we worked on our rhynchosaur numerous small boys would come to watch us, as well as grown-up passersby. All the while the traffic, such as it was, would rattle along on the nearby road. Most of the traffic consisted of two-wheeled carts, each pulled by a team of horses or of oxen, interspersed with heavy, solid-wheeled carts pulled by two, sometimes by three yokes of oxen. The oxcarts were frequently guided by a man on horseback or on foot, armed with a long pole and traveling along on one side, from which vantage point he could poke and guide the oxen as he saw fit. The big oxcarts were always heard from afar, because the axles were not overly supplied with grease. The drovers did this on purpose; they said that the noise of the wheels grinding on the axles comforted the oxen and made them work contentedly.

We labored to take out the rhynchosaur skeleton, using the methods that have already been described. Finally, when we had the blocks ready (for it required several) we flagged down a passing cart and made arrangements then and there to load the blocks and have them hauled to the Hübner place.

It wasn't as simple as that, because during all the time we worked at Quilômetro Tres—in Rio Grande do Sul for that matter—we were plagued by rains. My diary is full of references to work being interrupted by heavy, subtropical rains, which made our diggings muddy and which frequently resulted in the fossils on which we were working being surrounded by or even submerged in red, muddy water. Of course, when the sun came out and we waded back into our quarry to resume work we would find ourselves quickly covered with red mud and completely drenched with sweat as we labored in the hot, humid air. It is, however, the fate of the paleontologist to be more often than not uncomfortable while he is collecting fossils, so there was nothing to do but continue with our work, despite heat, humidity, and sticky mud. It was our work under such conditions that made the chimarrão breaks at the Perrera house and the rather long midday dinners at the Hübner place so very welcome. These were opportunities to refresh ourselves, to cool off after a fashion, and to replace lost body liquids.

During this time of frustrating rainstorms Margaret was scheduled to leave us, to return home. We had a family, or at least those of our boys who were still in high school, holding down our house in New Jersey. She felt she had been away long enough, even though her desire was to stay on through the duration of the expedition. On January 26, with black skies overhead, we all drove to the Santa Maria airport to see her off. As we waited in the airport we worried about our fossils getting drenched in the quarry; much more so, said Margaret, than about the prospect of her taking off into an ominous black sky. In due time the plane arrived, she was hurried out to it by an attendant carrying a large umbrella, and off it went into the blackness. I must say that I *was* worried about that, as well as about the fossils. Her journey proceeded on schedule; the plane flew through the storm to Pôrto Alegre, and from there she returned to New York.

We went back to our labors, and with diligent efforts in the Sanga de São José we eventually got our rhynchosaur skeleton, and a couple of others as well, which had been located after we had begun work on the initial specimen. Moreover, we explored some other nearby *sangas,* the Sanga da Matas and the Sanga de Baixo, where we located and collected still other skeletons, not only of rhynchosaurs but also of other reptiles as well, particularly some good-sized dicynodonts. These were bizarre therapsids or "mammal-like" reptiles in which the large skulls were decorated on each side with a big, downward-directed tusk. Except for the two tusks the dicynodonts commonly had no other teeth; rather, the jaws were beaklike, something like the jaws of turtles. In addition we found some pseudosuchian bones, representing small reptiles related in a general way to the crocodilians and to dinosaurs.

As a result of all this work we accumulated quite a pile of plastered blocks, which we stored temporarily in the Hübner barn. Then on February 15 we hauled in some lumber and proceeded to make boxes in which to pack our fossils. It was a busy day, but by evening we had completed the job, so that on the next day we were ready to load the boxes on a truck, for which arrangements had been made, to be hauled to Pôrto Alegre, from which place they would be shipped to their final destinations: some of them to Rio and some to New York.

It must not be thought that our time at Santa Maria was spent entirely with fossils. There were diversions. Such as the Sunday when I journeyed with Llew and Carlos and several other Brazilians to a fazenda west of the city. We drove out across the pampas, as birds by the thousands flew up and out of our path, and on the way we regaled ourselves with papagaio stories, having to do with parrots that talked knowingly about the doings of mere human beings. The day at the fazenda was given over to visiting and walking about and looking at horses and above all to feasting. It was a large family gathering, properly celebrated with mounds of delicious food ranged down the center of a long table.

Shortly before we were to leave Santa Maria, Carnival time came around—the great yearly celebration in Brazil that precedes Lent. The Carnival in Rio is famous; in Santa Maria we had a sort of miniature and less gaudy version of the affair. I was taken to a *clube* (a characteristic Brazilian institution) where people were in costume and there was much merrymaking. Finally I was worn out, so I retreated to the hotel, just down the street. But all through the night I was awakened by the sound of music and drums and shouts—sounds that continued until the sun appeared over the eastern horizon.

It was now time to shift our base of operations to the little town of Candelaria, about 100 kilometers to the east of Santa Maria. There we hoped to collect more fossils from the Santa Maria Formation, but fossils of a different kind. We were after a *fauna,* an association of animals that lived in Rio Grande do Sul many millions of years ago, when that land was very different from the pampas that one sees there today. We knew that we had only a part of a Triassic fauna among the fossils we had collected in the sangas at the edge of Santa Maria. Furthermore, we knew that other elements of the fauna might be found in the general vicinity of Candelaria.

The presence of different parts of a fauna at separate localities is not at all uncommon when one is involved with fossils. Indeed, it is common enough with modern animals. For example, if one is collecting mammals in the mountains of Colorado, a search may be made in the valleys for mule deer and coyotes and high on the mountain slopes for bighorn sheep and mountain "goats." Animals live in their preferred environments.

Such may have been the case in Triassic time, because it had been noted that whereas rhynchosaurs were found around Santa Maria, they were not found at Candelaria. There one might expect to find dicynodonts (a bit rare at Santa Maria) and particularly cynodonts, these latter being advanced, small to medium-sized mammal-like reptiles showing many of the anatomical features that were to characterize the mammals of later geologic ages. Perhaps the difference in the fossils found at Santa Maria and Candelaria is owing to differences in stratigraphic levels within the formation, but perhaps the difference is one of *facies,* reflecting a difference in environments. It has been thought by some paleontologists that the peculiar rhynchosaurs inhabited an environment separate from the habitat in which the cynodont reptiles lived.

However that may be, we were out to look for some cynodonts as well as other reptiles associated with them.

By this time we had acquired the use of a jeep, so we drove to Candelaria and established ourselves in the town's one hotel, a rather small building facing the little praça. Our host was the mayor of Candelaria. From this headquarters we worked the surrounding country, particularly some sangas near a little country store known as Pinheiros, because immediately in front of the building were two or three magnificent araucarian pines, the type of trees that flourished widely during Triassic times. (Araucarians are the trees that make up most of the fossil logs preserved in the Petrified Forest National Park in Arizona; today they are characteristic of southern continents.)

I tried to persuade my Brazilian friends that it would be to our advantage to take lunches with us, especially since Pinheiros is about twenty

Much of our work was carried on in the Sanga de Niconar, near Pinheiros, where from our quarry we could look to the north to the great escarpment of the Serra Geral.

kilometers from Candelaria. My arguments had no force whatsoever. Every morning, following a meager Brazilian breakfast of coffee, bread, and honey, we would drive to Pinheiros to spend a long morning searching for fossils or digging them out. The mornings seemed endless to me, because by about ten o'clock I would begin to feel acutely the lack of morning nourishment. Finally noon would arrive, and we would drive all the way back into Candelaria for a big dinner at the hotel. After that we were hardly physically prepared or in the frame of mind to venture out into the sun again, so we would have a long rest. Then in the late afternoon back to Pinheiros, where we might be lucky to get in two additional hours of work, since, as has been mentioned, darkness at this latitude comes on at about six o'clock. Well, that was the Brazilian way, and I had made up my mind that when in Brazil one should do as the Brazilians do.

Here, as in Santa Maria, we were plagued by rain, yet even so we unearthed some nice cynodont skulls and skeletons. Much of our work was carried on in the Sanga de Nicanor, near Pinheiros, where from our quarry site we could look to the north, to the great escarpment of the Serra Geral Lava which rises above the rolling lands of the pampas in the foreground.

The escarpment marks the southern edge of the plateau, one of the earth's great volcanic flows, formed during Cretaceous times when dinosaurs still ruled the land.

We worked in the bottom of the sanga, and there we found a natural little graveyard of cynodonts, which we collected. Our cynodont mine was productive, for after we had divided the fossils between us I was able to have eleven cynodonts shipped back to New York. We got some other fossils, too, especially dicynodonts.

The work of collecting need not be described. It was much the same story as at Santa Maria, searching and digging, with frequent rains to interrupt our efforts. When it was all over, we had some more boxes built in Candelaria, so that the fossils could be trucked to Pôrto Alegre.

As soon as the fossils were packed and loaded, Carlos, Fausto, and I drove back to Pôrto Alegre—Llew having already returned to Rio. And in due time I went back to Rio also, where again I was established in the hotel at Copacabana, where I could look out across the rolling waves of the South Atlantic.

Just a day or two before I was to take the plane back to New York I was awakened at about three o'clock one morning by a steady rhythmic drumming, coming in to me from the beach. I got up and went to the balcony, and there below me was a crowd of people facing the ocean and singing and beating drums, and before them were many candles. It was ob-

Gaúchos

viously a ceremony of some sort, and it continued for the better part of an hour.

At breakfast time I asked the clerk at the hotel desk about it, and he categorically denied that anything of the sort happened out on the beach in front of the hotel. I knew better, and later I talked with some of my Brazilian friends about it. What I had seen was a *Macumba* rite, performed to propitiate the gods. *Macumba* is a cult derived in part from West Africa, and mixed in Rio with spiritualism and with Catholicism to some degree. It involves various ceremonies, among them offerings of food to the deities. On the beach that night the worshippers had placed candles and food on the sands, where after a time the tide had advanced and then receded, washing the offerings out to sea. It was a primitive survival in a modern, sophisticated city, and evidently the hotel clerk did not wish to admit that such a thing could exist opposite *his* hotel. I should add that I saw the remnants of other *Macumba* offerings in Rio—bits of food and drink—one of them next to the Divisão where Llew Price worked.

With that little incident my first sojourn in Brazil neared its end. There were a few more details to be taken care of, and then on the evening of April 3 I boarded a plane at Galeão for my trip home. It was the end to a memorable visit, a visit to this other Eden, this tropical land that will always be affectionately emblazoned within my memory.

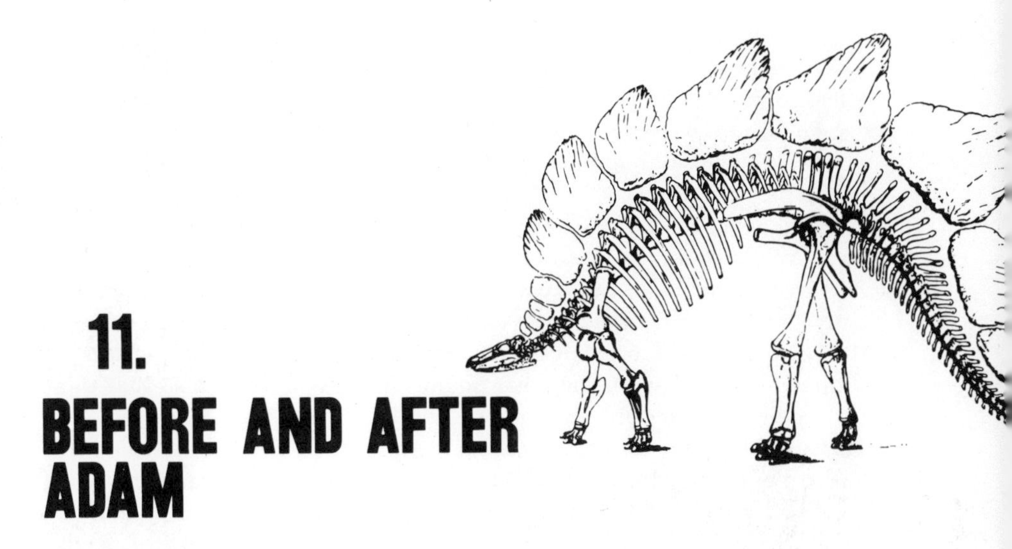

11. BEFORE AND AFTER ADAM

Several times in years past my very good friend George Haas of Hebrew University had urged me to visit Israel. I went there twice—once by myself in 1962, when I was on my way to Africa, and again with Margaret in 1965, following an autumn spent in southern Germany. Neither trip involved anything like a full-fledged collecting expedition, yet each was none the less professionally satisfying. Furthermore, we did collect some fossils.

Now for a few words about George Haas, a truly remarkable man. George is a big man physically and a most enjoyable person. When he ambles across the scene he always reminds me of a large and amiable bear. He is nearsighted, but he doesn't miss much. Many a happy day have I spent with George—in Israel and in the United States, which he has visited frequently. In some ways George might be considered as a sort of Renaissance man versed in many fields of knowledge. He is primarily a herpetologist and paleontologist, interested in living and fossil reptiles. But George has a remarkable comprehension of other fields of zoology, of botany, of the prehistory of the Middle East, and of history. He is a classical scholar and he speaks several languages. He also has a keen appreciation of art. Consequently a visit to Israel under George's auspices was a delightful experience in learning.

The first time I went to Jerusalem I found myself installed in a small hotel in the middle of the new city (Old Jerusalem was in those days still shut off from Israel by a formidable wall) where, on my first night of residence, I was kept awake until the small hours of the morning by the sounds of revelry emanating from a nightclub back of the hotel. It seemed strange, in this land, where biblical characters probably had walked on the ground beneath me, to hear a dance band playing "Tea for Two" through the night. The next day I changed to a room on the other side of the hotel, where there was plenty of traffic noise, but noise of the sort that was more tolerable for the sleeper than unwanted music. I must say, however, that the far from melodious calls (amplified by loudspeakers) of the muezzins across the wall, summoning the faithful to prayers, were less than welcome in the predawn hours.

Then began my introduction to Israel, from Metulla at the very northern tip of this small country, into the Negev on the south. I saw the northern extension of the great rift that traverses Africa from Lake Tanganyika northward through the Red Sea into the Gulf of Aqaba and on to the Dead Sea and the Sea of Galilee (or Lake Tiberias), where the land has dropped down between the great boundary faults of the rift so that these two bodies of water are, as everyone knows, considerably below sea level. Perhaps the people of antiquity did not realize this—not even the Romans, with all of their engineering skills. Subsequently, in 1965, near the shore of the Dead Sea Margaret and I saw the almost impregnable fortress of Masada, where from A.D. 70 to 73 Jewish patriots defied the might of Rome, finally to perish by their own hands, to and including the very last person who for three long years had occupied that stronghold on the frowning cliffs. At Tiberias we lunched on a terrace by the shore, with the uneasy

George Haas and Margaret at our hotel in Jerusalem. George might be considered as a sort of renaissance man.

feeling that the great guns on the Golan Heights on the other side of the lake were pointed in our direction. We visited Shivta, a long abandoned Byzantine city in the Negev, where a large population of people lived the good life in what is now a formidable desert. We saw the scenes of incredibly bloody conflicts between the Crusaders and the Saracens; at Acre and at the Horns of Hattin, where the Franks perished miserably beneath the weapons of Saladin's army, aided by a relentless sun. We visited the ancient Roman city of Caesarea, subsequently transformed into a medieval fortified port. We saw the steplike exposures of Cretaceous limestones around Jerusalem, which for thousands of years have been converted into terraced fields and vineyards. At many more places we saw history through the millennia when many peoples marched and countermarched through this crossroads of the Middle East. And we almost became the unwilling spectators of a small bit of contemporary history, because one day along the Jordanian border a group of Arab guerrillas crossed into Israel just where we had been, and a sharp little fight ensued. That was a few minutes after we had gone on with George Haas to see something else.

Of history there is a wealth almost beyond imagining in this land bounding the eastern curve of the Mediterranean, but there is prehistory too, from the days of early man back to the days of dinosaurs and beyond. That was what I particularly wanted to see.

Makhtesh Ramon is a great depression in the desert south of Beer Sheva (known to readers of the King James version of the Bible as Beersheba), an elongated east to west topographic basin some twenty-five miles in length, and perhaps four or five miles across. In geological terms it is an eroded anticline, which means that it was formed by an upward warping or folding of the crustal strata, accompanied and followed by erosion along the axis of the fold (the zone of greatest weakness) to create the basinlike depression. Thus the rocks forming the rim of the basin are the youngest, being Cretaceous in age—the last period of the Mesozoic—while those in the middle of the basin are the oldest, in this case Jurassic and Triassic, representing respectively the middle and early Mesozoic periods. The depression is not quite so simple, for its southern boundary is marked by a long fault that was formed by tensions that developed during the anticlinal folding. Needless to say, the Triassic beds in the middle of the depression were the center of attention for both George Haas and myself.

From Mizpe Ramon on the northern rim of the depression, Makhtesh Ramon is indeed an impressive sight. The floor, far below the rim, is a rugged desert terrain, formed of eroded brown and gray and sometimes red badlands, traversed by the courses of perennial desert washes, and startlingly punctuated by black mounds, the remains of past volcanic activity. Far away on the other side is the long, level rim, a steep escarpment of light-colored limestones and sandstones.

Makhtesh Ramon is a great depression in the desert south of Beer Sheva. The Triassic beds in the middle of the depression were the center of attention for both George Haas and myself.

We had stopped here for an extended view of Makhtesh Ramon, and having seen its awesome starkness we climbed back into the Land Rover for our descent into a landscape that already by midmorning was shimmering in the heat. "We" consisted of George, two of his advanced graduate students, Eviator Nevo and Eitan Tchernov (both of whom have earned enviable scientific reputations), and myself. It was to be a hot day down in the depression, but that is a common situation for the fossil-hunter. Indeed, I could almost imagine myself back in the Arizona desert—north of Cameron, for instance.

The sediments in Makhtesh Ramon are primarily of marine origin, so the objects of our quest were the bones of Triassic reptiles that had lived here when this was a shallow seaway. We found them, particularly the remains of placodonts, which were strange armored reptiles, rather turtlelike in shape (yet the placodonts were not related to the turtles) with broad, crushing teeth in the jaws by means of which they were able to crack the shells of various molluscs that lived here in profusion. Such was their diet. We did not attempt to collect any of the placodonts, although some of the better ones were marked for future investigation.

We did have a good time, even though a hot one, clambering over the steep little badlands and among the sharp, eroded rocks that make the floor

On this particular outcropping the entire surface was a mass of large ammonites. Eviator Nevo (left) and Eitan Tchernov (right), at the time graduate students, are now professors and acclaimed research workers—Eviator at Haifa, Eitan at Jerusalem.

of Makhtesh Ramon. Late in the afternoon we crossed the southern boundary fault and came to one of the most spectacular fossil outcroppings I have ever seen. It was an upturned wall of Cretaceous limestone paralleling the fault and replete with the coiled shells of an ammonite known as *Leoniceras*. (The ammonites were Mesozoic cephalopods related to the modern pearly nautilus, and they lived in remarkable variety and profusion.) On this particular outcropping the entire surface was a mass of large ammonites, many of them as much as two feet or more in diameter. What was the reason for the existence of this ammonite burial ground? Why had so many of the ancient "shellfish" perished, to be preserved as fossils at this particular locality? It was a subject for speculation.

Postscript. Three years later, when Margaret and I were in Israel together, we returned with George to Makhtesh Ramon and collected a nice placodont for New York. And while waiting for the plaster to dry on the specimen being collected we poked about and found two more of these strange reptiles. Also, while all of this was going on Margaret surreptitiously smoothed a gob of surplus plaster that had been discarded and on it she carved an inscription: "Here lies a long-placated placodont." She buried it near our excavation, where it was found two years later by George—much to his amusement and delight.

Of the Middle East Triassic I saw as much as I could, and what I saw linked the badland exposures down in the lowest levels of the Makhtesh Ramon, in a terrain devoid of vegetation and brutally hot beneath the midday sun, with the Middle Triassic Muschelkalk limestones seen in the cool, green forests of southern Germany. It was the record of a time, long ago, when armored placodonts, long-necked nothosaurs (the precursors of the plesiosaurs) and other reptiles, as well as many fishes, swam in the tropical waters of the Muschelkalk seas. There was another glimpse into the Age of Dinosaurs, this time a rocky floor uncovered on the outskirts of Jerusalem on which are exposed hundreds of footprints of a rather small dinosaur, of the kind known as ornithomimid. These were the birdlike dinosaurs which in Cretaceous times roamed across the landscape on long legs in pursuit of small game or in escape from their enemies. The tracks made a good show, and have since been protected as a display for the public, with a life-size model of the type of dinosaur to which they belonged. It is always nice to see dinosaurs getting the recognition that they deserve.

In Israel the records of prehistory that are of particular interest to large numbers of people are those pertaining to early man and his environment, including the animals and plants that were his contemporaries. One of our casual contacts with early man in the Middle East was the day when we were returning to Jerusalem from Makhtesh Ramon and got thoroughly stuck in a desert wadi. As we dug around the car in an effort to free it from the seemingly bottomless sand, Margaret wandered off down the wadi to see what she could see. What she saw was a beautiful flint ax protruding from a bank. She plucked it from its resting place and brought it back to the scene of our troubles, and immediately George identified it as a Mousterian artifact. Here was a record of Neanderthal man in the Middle East at a time when it was the locale for hunters rather than warriors. It was the warriors who got us out of our predicament, for the Israeli Army was on maneuvers in the Negev, and eventually a powerful desert vehicle came to our rescue.

A less casual visit to Neanderthal man was to be our good fortune a week or so later. On that occasion we journeyed north from Jerusalem, and during the course of our travels we one day visited some caves in cliffs at Mount Carmel, near the coast at a locality known as Ma'agan Mikhael, about twenty miles south of Haifa. As usual George Haas was our more than perceptive guide, and he was accompanied by Eitan Tchernov. Along the way we were joined by Professor Raymond Dart of South Africa and his wife and by Professor Stekhelis of Hebrew University and his wife. Dart was the original describer of *Australopithecus,* the ancient humanoid that has since become pivotal in the studies of early man; Stekhelis was an archaeologist noted for his studies of Middle East prehistory.

We reached the caves, and in particular a cave where Stekhelis and

Haas had been working, a cave where a Neanderthal child had been found, and where there was discovered a buried hyena skull surrounded by a ring of stones. Could there have been a "hyena cult" among the Middle East Neanderthalers, comparable to the "bear cult" burials in Europe, where skulls of the huge Cave Bear were religiously interred? Perhaps.

It was cool in the cave, which had a high roof where there was a constant fluttering of bats, and we perhaps felt reluctant to emerge into the hot sun. In front of us, between the cave and the sea, was a small flat plain, covered by a banana grove, so we retreated beneath the banana trees for our lunch, and there we feasted most delightedly on some fresh bananas still on the trees, evidently left there after the harvest. Perhaps they had not been sufficiently ripe at the time. They certainly were perfection for us; we pronounced them the best bananas we had ever eaten.

Before our visit to the caves and the banana patch we had all been to a locality along the Jordan River known as 'Ubeidiya, where Hebrew University had been excavating on a rather large scale, right at the edge of a vineyard. No, we did not eat any of the grapes, but we did see long, deep trenches, and what was coming out of them. It was a rich and impressive fauna—mammoths, ancient horses, rhinoceroses, pigs, hippopotamuses, giraffes, antelopes, saber-toothed cats, lynxes, hyenas, bears, wolves, numerous rodents, turtles, crocodiles, and birds. With these ancient animals were primitive stone tools, comparable to the artifacts found at the famous Olduvai site in eastern Africa. Here was a view of life in the Middle East long before the days of the Neanderthals. When we were there a few human teeth had been found, and some of the participants in the dig were speculating as to their possible relationships.

Down along the western shore of the Sea of Galilee is Capernaum, where Christ talked with the Roman centurion, and while in that vicinity we visited the mosaic floor of the loaves and fishes—the still visible remains of what was once a chapel. This exquisite and large mosaic is now protected from the weather by a roof. While we were looking at it two nuns came in and, seating themselves at one end of the building, began to sing most sweetly. The songs were ancient airs, and Eitan, who among other things is a musicologist, identified them for us. He was fascinated by the singing, as we all were. It was a memorable experience that has stayed with all of us through these years; I know because Margaret and I speak of it, and in a letter not long ago Eitan mentioned it.

A last word about early man and animals in Israel. One day during our journeys we visited a kibbutz (Sha'ar HaGolan) to the south of the Sea of Galilee and only a kilometer or two away from the Syrian border. The Syrians had the unpleasant habit of dropping artillery shells from the Golan Heights into the settlement on odd occasions, so the people of the kibbutz had dug some bomb shelters and trenches to which they could retreat for protection. In the process of digging a deep shelter they had turned up an archeological site, with bones of men and animals, and artifacts. What did they do but make the underground shelter into an underground museum as well. Around the walls were exhibition cases, displaying the bones and artifacts found at the site, very well labeled. Indeed, they had obtained the help from university archeologists and zoologists, so that the objects were authoritatively described. It was so effective as a local display in place that classes from the schools in Tiberias and other towns came to the kibbutz to see the little museum. A class arrived while we were there. For us it was an object lesson of fortitude and ingenuity; it epitomized the spirit of a people who make the best of their opportunities and refuse to feel sorry for themselves.

We were entertained with an excellent high tea at the kibbutz and then we went on our way, to other places where the prehistory of this land could be seen, and eventually back to Jerusalem. For us that was Israel, a land that has been called holy.

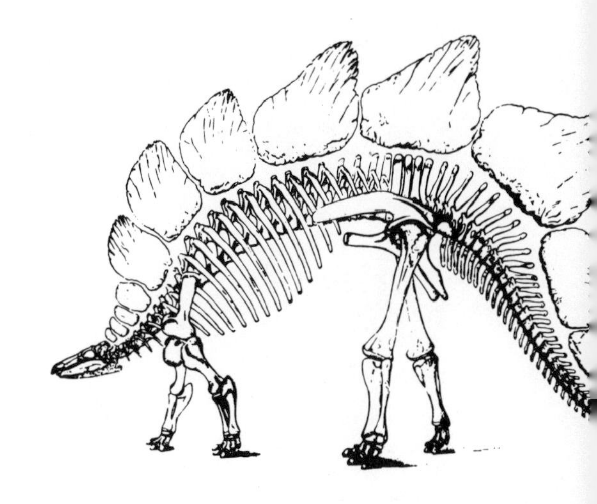

12. VELDT AND DRAKENSBERG

As I left Israel on June 14, 1962, after my first visit to that historic land, my course was to the south, and no matter the speed of the plane, my thoughts ran far ahead to the land toward which I was flying. The land was Africa and my thoughts were of Africa, but not merely of Africa as we know it. I was thinking of Africa as it was in the distant past, of Africa during Permian and Triassic times, when it was a part of the great supercontinent of Gondwanaland—at least, according to the theories of plate tectonics and continental drift.

The theory of continental drift was to a large degree the brainchild of the German meteorologist Alfred Wegener, and it came into being about 1912, when I was still a very small boy. The theory proposes that at one time in the past all of the continents as we know them were joined to form a single supercontinent, which has been called Pangaea. This great land mass presumably was formed of two moieties, a northern hemisphere part consisting of what is now North America and most of Asia, and a southern hemisphere part comprising South America, Africa, peninsular India, Australia, and Antarctica. The former has been called Laurasia, the latter Gondwanaland. At some date in the past, possibly during the final years of Triassic history, the great supercontinent began to split apart—a process that was many millions of years in the making—and in time the several fragments of this rifted land drifted in various directions to occupy, as the continents in their present forms, the positions where they are now situated.

It was the study of Gondwanaland that gave much of the impetus to the early work on continental drift. Wegener was a vocal and tireless advocate of the new theory, and he was seconded at an early date by the South

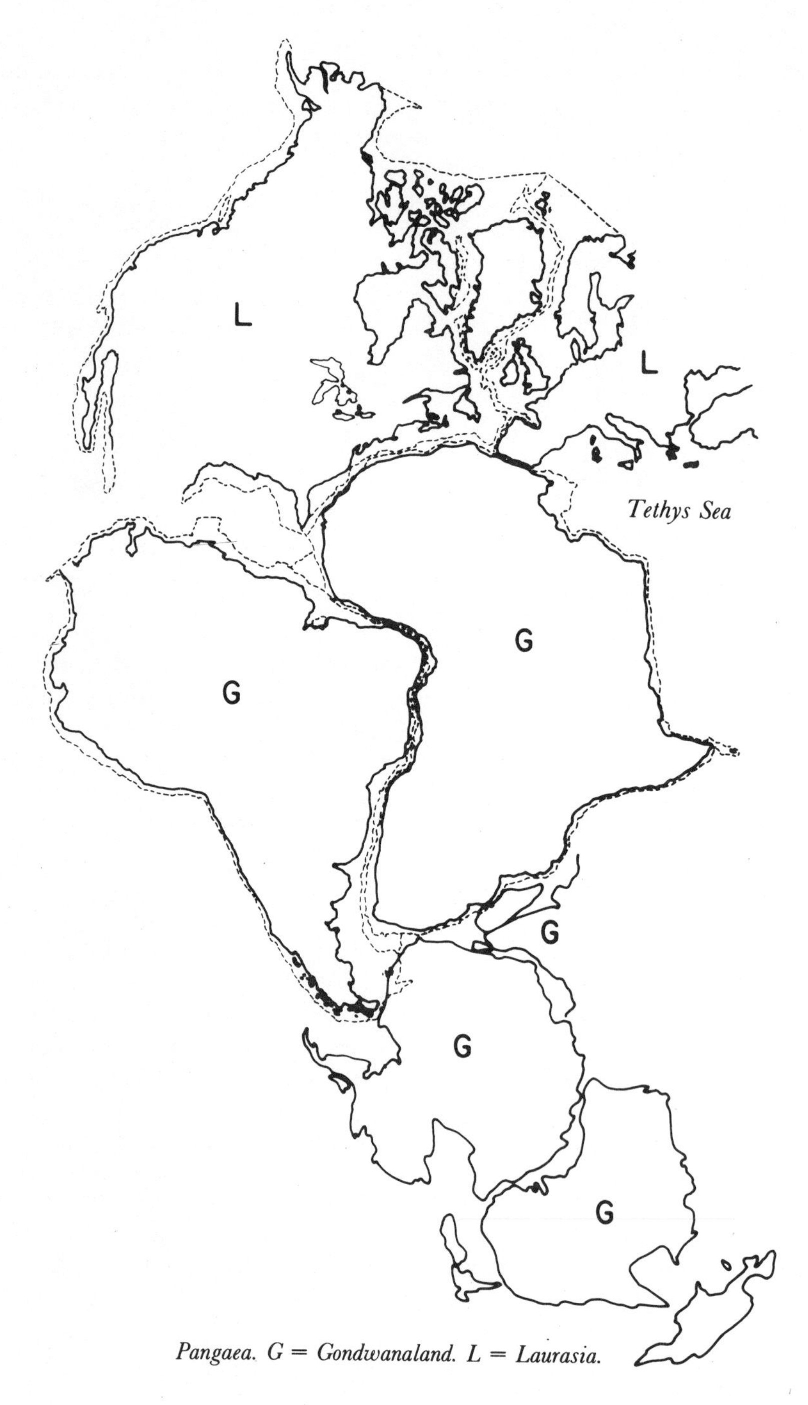

Pangaea. G = Gondwanaland. L = Laurasia.

African geologist Alex du Toit.

In the beginning it was not a popular theory, especially among geologists and biologists in the northern hemisphere. The southern hemisphere scientists were on the whole more sympathetic to this new idea. Consequently, Wegener and du Toit stood pretty much alone and were regarded by many of their colleagues as chasing a scientific will-o'-the-wisp. The trouble was that the continental drift theory was founded on a small amount of evidence, and for the most part evidence that could be used just as legitimately to explain the earth in the terms of fixed continents and ocean basins. Indeed, the celebrated American geologist Bailey Willis in 1932 stated emphatically the dictum that expressed very nicely the overriding concept among most geologists: *"Once a continent, always a continent. Once a basin, always a basin."*

Then came the Second World War and the development of techniques, especially sonar or echo-location methods, that could be applied to the mapping of ocean basins. Soon it became evident that the oceanic basins are quite unlike the continental regions. Other techniques also were developed, especially in the field of geophysics, whereby, among other things, the ancient magnetism of the earth could be measured. Within a relatively few years—a decade or so—it began to seem as if Wegener and du Toit had been right after all. They had been prophets ahead of their time, relying upon their convictions even though the hard evidence to back up their ideas was largely not at hand. In this manner the modern concept of plate tectonics came into being, the result of work by many authorities working in various fields of scientific endeavor.

Plate tectonics (which involves continental drift) is an elegant concept, and so much corroborative evidence accumulates year after year in its support that the idea is no longer in the realm of pure theory; it has become a fact of scientific life, just as Darwinian evolution has become a fact of scientific life. As the development of the theory of evolution a century ago brought about a great revolution in our view of life, so the development of the theory of plate tectonics has brought about an equally significant revolution in our view of the earth. The earlier theory, largely the work of one man, Charles Darwin, changed forever our concept of life as preordained and static, to replace it with a view of life as constantly changing and evolving. The newer theory , the work of many men, has revised our concept of the earth from that of a static globe to that of a mobile and ever-changing sphere.

I had long been very skeptical as to the reality of continental drift, because it seemed to me that the distributions of land-living animals through time could be explained in terms of fixed continents. In this I was of course echoing the views of many scientific workers of my generation—people involved in geological and biological studies. As my plane flew across

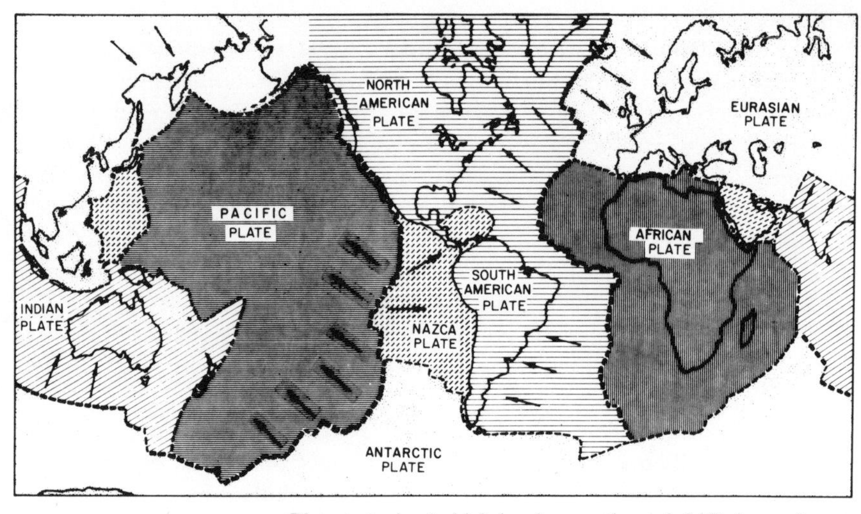

Plate tectonics (which involves continental drift) is an elegant concept, no longer in the realm of pure theory; it has become a fact of scientific life. It has revised our concept of the earth from that of a static globe to that of a mobile and ever-changing sphere. This shows the major and minor plates and their respective directions of movement.

the starkly beautiful Persian desert toward Kenya I was thinking at times about Africa and about continental drift. We flew over Iran because I was in an El Al plane, which was prohibited from flying over unfriendly Arab countries. So we went out over the Mediterranean after we left Lyd airport, circled north through Turkey, then down into Iran, and from there over the Arabian Sea to Nairobi. It was the long way around, but it made for an interesting and beautiful flight.

Plate tectonics and continental drift are all very fine, but would the evidence of the fossils accord with this ever more popular and ever more accepted view of our earth? As David Elliot, one of my future very good friends and colleagues, said: "Geophysical measurements and related studies are convincing, but nothing can dispute the authenticity of a fossil bone with a valid record as to its occurrence." Or words to that effect. David's dictum was very comforting to me, coming as it did from a "hard-rock" geologist who has devoted his life largely to the study of volcanic rocks.

Here I was, on my way to Africa—the central core, one might say, of ancient Gondwanaland. What was I to see? Certainly what I was to see would be viewed against a background of plate tectonics and drift, and would the background be discordant or harmonious with the fossils? When I had been in Brazil I had not been thinking so much about Gondwanaland as the possible locale for the fossils we found; now I was.

In a sense I was going through still one more change of direction in my scientific life—perhaps not so much a change of direction as a change of emphasis. I was going from the more or less strictly paleontological study of ancient reptiles to work on these fossils against the background of their former distributions. Perhaps it was my previous shift from mammals to reptiles that in time brought about an increased interest in ancient distributions as related to the problem of plate tectonics and drift. Certainly the reptiles of the late Paleozoic and the Mesozoic eras are in many respects more crucial to the implications of drift than are the mammals, especially the later mammals, which lived at times when the continents were assuming their present positions and connections.

Africa was, in many respects, the center of the Gondwana continent, the center that perhaps remained more or less in a stable position while the other fragments of the ancient continent broke away from the African core and drifted in their several directions; South America to the west, peninsular India to the northeast, Australia to the east, and Antarctica to the south. I was on my way to the heart of Gondwanaland.

As the plane comes in over Johannesburg one cannot but be impressed by the tremendous mine dumps that edge the city. They are huge almost beyond belief; great square truncated pyramids that dwarf the buildings and the roads and the cars and the people that surround them. They symbolize in a way one of the many contradictions of Africa; in this case the superimposition of massive engineering works introduced by European man on a land recently pristine and a culture retaining many of the attributes of an earlier age.

The contradiction became evident once I was down on the ground. At the edge of the modern airport I boarded a bus, and for the few moments before it departed for downtown Johannesburg I watched black people, descendants of early inhabitants of this land, building little fires in front of makeshift shelters on some vacant ground in an effort to keep themselves warm. In spite of the sunshine it was cold, for this was June—the middle of winter in the southern hemisphere.

At the terminal in town I was met by Ian Brink, then the director of the Bernard Price Institute for Paleontological Research at Witwatersrand University, and James Kitching, the senior paleontologist at that institute. I had known Brink from former years, when he spent some time as a visiting scientist in the United States, but this was my first meeting with Kitching. Although I could not know it at the time, the scientific lives of Kitching and myself were destined to be closely linked in years to come. We had a pleasant meeting together, and then I went with Ian to his home for a reunion with his wife, Anna, and his family, and a few days at the university. These days were the prelude to a long-anticipated adventure awaiting me, a trip through the South African Karroo with James Kitching.

To the paleontologist, especially the vertebrate paleontologist, and most particularly the vertebrate paleontologist interested in the evolution of the mammal-like reptiles, the Karroo is a fossil field beyond comparison. There is no place in the world like it.

The Great Karroo is an elongated geological basin occupying the southern tip of Africa, its western extremity being about 90 or 100 miles from the western coast above Cape Town, its northeastern end extending beyond Johannesburg to a similar distance from the eastern coast above Durban, its northwestern border paralleling for some distance the course of the Vaal River, and its opposite, southeastern border truncated by the edge of the continent bordering the Indian Ocean. It is a large basin, 700 or 800 miles in length from southwest to northeast, and some 400 miles across. Since it is so large, the traveler who crosses the Karroo is hardly aware of its structure as a geological basin, for topographically it is a vast desert or semi-desert, its surface intermittently covered with buttes, mesas, and tablelands.

It may be remembered that Makhtesh Ramon in Israel is a topographic basin developed on an upwarped structure—a large eroded anticline with the youngest geological formations around its periphery, the oldest in the center. The Great Karroo, being a structural basin, shows the opposite relationships, with the oldest formations around its edges and the

A characteristic Karroo mesa, the sloping sides formed by permo-Triassic sediments, the steepcliffed cap by ancient volcanic rocks. The Karroo is a vast desert or semi-desert, its surface intermittently covered with buttes, mesas, and table-lands.

youngest ones in the middle. The focal point of this great basin is occupied by the rugged, volcanic Drakensbergs, rising to heights of 10,000 feet or more above sea level. These high mountains form a sort of high island near one end of the Karroo, for the basin is somewhat tilted and thus asymmetrical. Consequently, the Great Karroo consists of a series of concentric sedimentary rings around the Drakensberg core, progressively older as one journeys from the Drakensbergs to the edge of the basin. As has been said, the Karroo Basin is truncated by the edge of the continent along its southeastern edge, and this truncation has cut the outer sedimentary rings of the structure so that they are here incomplete. The Karroo Basin is bounded on its southwestern part by the folded mountains of the Cape; thus one has to traverse narrow and picturesque passes if the Karroo is to be entered from Cape Town. But from Johannesburg the journey into the Karroo is less abrupt; one travels along roads that lead from the bustle of a great metropolitan region into the ever more quiet and more rural reaches of the African countryside.

That was how Dr. Kitching and I began our reconnaissance trip through the Karroo. It was a lovely land through which we traveled, a land of farms and villages, of cattle grazing across the winter fields and everywhere of rocky slopes and cliffs rising to terminate in flat-topped plateaus. We drove through Heidelberg, a typical little South African town with its church in the middle of the square, and on south to Harrismith, where we arrived in the late afternoon and put up at a local hotel.

We had an hour or two before dinnertime so we drove outside of town to explore some interesting-looking exposures. Our little trip immediately bore paleontological fruit in abundance, because in the course of a half-hour or so we found six specimens of the Lower Triassic reptile *Lystrosaurus,* the characteristic fossil of the Middle Beaufort beds of South Africa. It was my first introduction to *Lystrosaurus* in the field, and it was the beginning of an association with this fossil reptile that was to be of great consequence in my life. Before going any farther, let us see where *Lystrosaurus* fits into the sedimentary scheme of things in the South African Karroo.

The Upper Permian and Lower Triassic strata of South Africa have been classified in the manner shown in the chart, with, as is usual in geological practice, the youngest formations at the top of the sequence and the oldest ones at the bottom. (That, after all, is the way they occur in the field.)

This sequence of sedimentary rocks forms one of the finest records of vertebrate evolution to be found in the world. The associations of fossils—the faunas—are characteristic of their respective levels, and they demonstrate in a most elegant manner the evolution of their constituent members as one goes up in the sequence from the older to the younger rocks. Here one sees the development of life through time—the evolution of nu-

LOWER JURASSIC	Stormberg Series	Drakensberg volcanics
UPPER TRIASSIC		Cave Sandstones Red Beds Molteno Formation
LOWER TRIASSIC	Beaufort Series	Upper Beaufort — *Cynognathus* Zone Middle Beaufort — *Lystrosaurus* Zone *Daptocephalus* Zone
UPPER PERMIAN		Lower Beaufort — *Cistecephalus* Zone *Tapinocephalus* Zone

merous animals and the changing complexion of the faunas to which they belong as a result of this evolution.

As can be seen, *Lystrosaurus* is about in the middle of things. This strange-looking reptile, quite distinctive in the structure of its skull, was a squat, heavy animal, ranging in size from individuals about comparable to small dogs to those as large as sheep, or even larger. The body was robust and barrel-like, the limbs were strong, and the feet were broad. The tail was short. The skull, rather flat on top, had a down-turned muzzle in which there was on each side a large tusk. Otherwise there were no teeth in the jaws, which were developed as a sort of beak, not unlike the beak of a turtle, and in life were certainly encased in horny sheaths. The eyes were situated high in the skull, as were the nostrils. Obviously *Lystrosaurus* was an herbivorous reptile (as indicated by the capacious body, adapted to contain an elongated digestive tract for the processing of plant food) that lived in and around water (as indicated by the highly placed eyes and nostrils). Moreover, *Lystrosaurus* is by far the most abundant reptile in the Middle Beaufort beds, another indication it was a plant feeder that formed the base of a "food pyramid" upon which rested the ecological structure of many other reptiles with which it was contemporaneous.

Because *Lystrosaurus* is so abundant and distinctive it makes a very useful "index fossil" for the identification of basal Triassic sediments, wherever it occurs. As is evident from the names of the several zones in the Beaufort beds, other distinctive reptiles serve as index fossils for the levels at which they are found.

Lystrosaurus is a therapsid or mammal-like reptile, albeit a very highly specialized member of the group. Among the mammal-like reptiles it belongs to a very large group known as dicynodonts, a name indicative of the two tusks so commonly characteristic of the group. Taken as a whole, the therapsids are those reptiles intermediate between the reptiles and the mammals, and they are particularly abundant in the Beaufort Series of the Karroo. (The mammal-like reptiles decrease dramatically with the advent of the Stormberg Series, where those reptiles that were to rule through the remainder of the Mesozoic era, notably the dinosaurs, become domi-

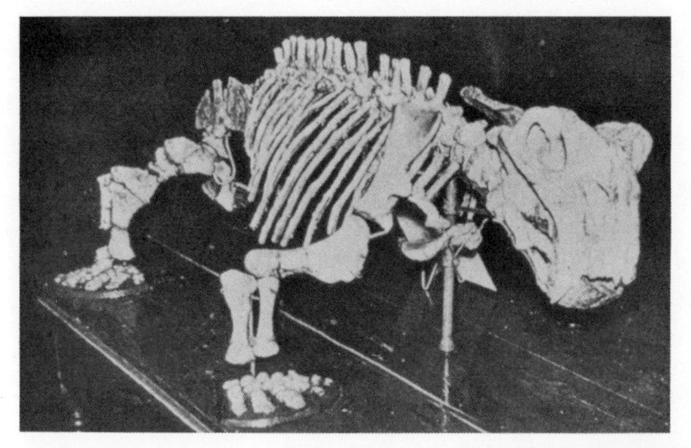

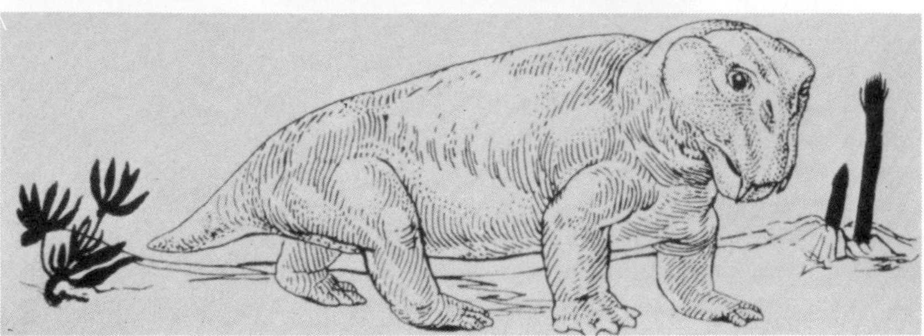

Lystrosaurus, *a strange-looking reptile, quite distinctive in the structure of its skull, was a squat, heavy animal. (Skeleton from a photograph by A. W. Crompton. Restoration by Margaret Colbert; from Edwin H. Colbert,* The Age of Reptiles, *Weidenfeld and Nicolson, London. W. W. Norton, New York.)*

nant.) The dicynodonts were specialized in a very definite direction and are not situated on the direct line leading to the mammals. Some other mammal-like reptiles are, however, very mammalian in their anatomy—so much so that in certain groups of them it is truly difficult to be sure whether they were still on the reptilian side of the threshold between the two great groups of vertebrates, or whether they had already crossed over to become mammals of a very primitive sort.

This digression will perhaps give some background for a better understanding of our excursion in South Africa, and beyond that, for paleontological work that will be described in subsequent chapters.

That first afternoon on the outskirts of Harrismith set the pace for our little trip in the Karroo, for it seemed that everywhere we went we found fossils. All of which is some indication as to the abundance of fossil reptiles in the Karroo beds. I have never seen anything to equal the numbers of

Our success in finding Karroo fossils was not entirely a matter of their abundance; much of it depended on the sharp eyes of James Kitching.

fossil vertebrates in the Karroo, except perhaps the prolific occurrences of Oligocene mammals in the White River Badlands of South Dakota. Wherever one goes in the Karroo there is a feeling of fossil reptiles at one's feet—and more often than not the fossils *are* nearby, waiting to be found and dug up. Moreover, the excavation of Karroo fossils is often facilitated by the fact that they frequently occur in concretionary nodules, so it is necessary only to stoop down and pick them up. The removal of a Karroo reptile skull from its encasing matrix can be something else, too often a matter of weeks or months of tedious work in the laboratory.

I must admit that our success in finding Karroo fossils was not entirely a matter of their abundance; much of it depended on the sharp eye of James Kitching. I have worked in the field with many bone hunters, but never have I seen anyone who could spot fossils as can James. His aptitude in this respect is truly uncanny, and sometimes frustrating to his field companions. All of us dearly love to find fossils in the ground, but when one is with James and he finds nine fossils for every specimen that you yourself can see—well, what is to be done? Nothing, except to plug along over the ground in silent envy and admiration. It did not take long for me to appreciate how fortunate I was to be conducted through the Karroo by this masterful fossil-hunter.

Never can I forget the winter mornings in the Karroo. Out into the crisp, cold air after breakfast, with the slanting rays of the early morning sun reflected as thousands of bright diamond points from the wet grass, and the fragrant smoke from the morning fires in the African houses hanging in a low, still cloud over the settlement, situated a bit apart from the White town. The shouts of the black drovers as they urge their oxen along the roads. And from all sides the gentle notes of the doves—to me one of the very characteristic sounds of southern Africa. There was magic in this scene, a magic that enhanced the beauty of the landscape, the beauty of a land where one could wish that there were not such overriding problems of human relationships.

(At this place let me say that I was and am keenly aware of the problems in this troubled land, but I do not choose to expand upon them at this place. They have been discussed in great detail and from many points of view by numerous authors, qualified and less qualified, and the conclusions reached have been legion. Suffice it to say that the facts of human relationships in southern Africa are incredibly complex, and suffice it to say also that I am writing about paleontological experiences. I was in Africa on grant money for scientific purposes; it was not my place or my intention to get involved in social problems. I kept my eyes and my ears open and came to my own private conclusions.)

From Harrismith, James and I drove on to the Golden Gate, a magnificent series of cliffs in the Red Beds and the Cave Sandstone, the latter so named because these hard, fine-grained sandstones weather into pinnacles, overhung caves, and undercut ledges, thus making rock shelters that have long been used by people, prehistoric and modern, for protection against the elements. We then continued to the town of Senekal, where the town square is completely bordered by large fossil logs of Triassic age, laid end to end.

There is no point in recounting the details of our Karroo trip, a trip where day after day we studied the sequence of the sedimentary rocks and tramped over the eroded slopes looking for fossils. They were invigorating days, made lively by the discoveries of mammal-like reptiles and by the numerous little adventures that are a part of the fossil-hunter's life. We ranged up and down through the Beaufort Series in search of fossils characteristic of the several levels—and we found them.

S. H. Rubidge was a prosperous farmer who owned a large acreage not far from the town of Graaf-Reinet. A good many years ago he had found some fossils on his land and, being curious as to what they were, got in touch with Dr. Robert Broom, the famous gifted and eccentric South African paleontologist. He sent some fossils to Broom, expecting a letter giving identifications of the specimens and some information as to their significance. Instead, Broom appeared on the scene, and thus began an association of considerable paleontological import. Mr. Rubidge became an enthusiastic fossil collector; he found fossils and he saved them. Moreover, he made them available to Broom and other paleontologists for study. But he did want the fossils to stay on the land where they had been discovered, so he built a museum building near his house, and there he displayed a large and important collection of Karroo reptiles. James had known Mr. Rubidge for many years, in part because James in his young life had been a special protégé of Broom. He therefore derived much pleasure in taking me to the Rubidge farm, Wellwood, to meet the family and to see the fossil collection.

Wellwood is a large, almost patriarchical establishment, a well-proportioned and very ample house of Cape architecture, with the usual barns and other outbuildings, and back behind a row of neat cottages for the African workers. There were clipped lawns and arbors, the grass crisp under the winter sun, and a contingent of very impressive rams. Inside the house were elegant furniture and well-stocked bookshelves and many amenities for comfortable living. James and I were served with a delicious midmorning tea, and then we went to look at the fossils in the museum. After that Mr. Rubidge took me for a little walk around the place, during the course of which he waxed warm about world politics and the injustices done to South Africa. But I kept my peace and admired the surrounding landscape.

Then James and I went to Graaf-Reinet, and on to Doornplaats, the farm of Mr. and Mrs. Pringle Brodie, where we stayed for several days. The Brodie house, a plain structure with a corrugated iron roof (as is so prevalent in the Karroo), had a deep verandah supported by six pillars across the front of the building—a pleasant place to sit on a sunny day. From this comfortable spot one looked out across the veldt, where a mile or so away a large hill or butte rose stark against the blue sky. Its slopes, sparsely covered with short grass, were formed by flat-lying Beaufort beds belonging to the *Daptocephalus* Zone, and it was capped by a horizontal layer of volcanic dolerite that formed a palisaded cliff, where it rose abruptly from the sedimentary slopes. On the near side of the hill a volcanic dike formed a straight wall, cutting through the sediments and rising from the lower part of the hill toward the flat, volcanic cap. It was a characteristic Karroo physiographic feature, so nicely expressed that I took a picture of it and sketched it as well. Little did I realize at the time that I was looking not only at a Karroo hill but also at a bit of Antarctica—without the snow and ice. More of this later.

The Brodies, of English antecedents, were our generous hosts, entertaining us, dining us, and giving all possible assistance to our fossil collecting. The assistance was much needed and appreciated, because one day James (of course) found a fine skull of *Daptocephalus,* a very large dicynodont reptile.

It was not the kind of fossil that one can pick up and put in a haversack. Therefore we made arrangements to go and get it, with the help of

Mr. Brodie's son, Mark, and some of the African employees. Early on the morning after James had made his discovery, we all rolled out of our warm beds, and as the sun rose over the flat-topped butte that faced the farmhouse we piled into a two-wheeled cart pulled by a tractor and bounced over the rugged terrain to the fossil site. One member of our party was Bess, the Pringle bitch, who was suffering more than a little bit from a sore paw that she had cut on a sharp stone a few days previously. But would Bess be left behind—not a bit of it. She insisted on being right in the middle of the things, and when we got out of the car to climb a long slope to the skull, she painfully limped along with us. As we climbed we flushed a herd of rhebok, one of the African antelopes—elegant little animals with sharp, slender horns.

It was a heavy specimen but, like so many Karroo fossils, very solid. Consequently, we did not need to apply any of the hardening and plaster techniques that have been described in order to collect the fossil. It was mainly a matter of muscle power, and we had that. The skull was successfully pried loose from its resting place and raised into the cart. Then back to the farmhouse, a most welcome warm breakfast, and a good feeling for having secured such a fine fossil.

Perhaps that was a high point of our trip. We reluctantly left the Brodie family and went on, and four days later were in Cape Town in good

Little did I realize at the time that I was looking not only at a Karroo hill but also at a bit of Antarctica—without the snow and, ice.

time for the meetings of the South African Association for the Advancement of Science. Margaret flew in, just before the meetings were to open, and like many of the delegates we were housed in the dormitories of the university. Perhaps it was appropriate that our first night there was interrupted by the roaring of lions (they were in the zoo), but it was pleasing in between roars to hear the musical, bell-like calling of thousands of frogs that had come out on the wet lawns of the campus.

The meetings went according to schedule, including an all-day trip on the Fourth of July to Stellenbosch, where we were served an enormous lunch, hosted by the city fathers, after which Margaret and I rested at the home of one of the professors. There we admired the folded mountains of the Cape, rising in pastel splendor back of the town. In the evening we all attended a *braaivleis,* the African version of a barbecue, and there I chanced to sit next to the manager of one of the big wineries, one of the establishments producing fine South African wine for export. "Don't worry about what kind of wine to drink," he said. "If you like white wine with meat go ahead and drink it, or red wine with fish. Suit yourself." Obviously he was no wine snob.

Now we got ready for another African adventure. James Kitching had led me through the Permian and Lower Triassic Beaufort beds; now I was to see the rocks of the Upper Triassic Stormberg Series under the tutelage of Dr. A. W. ("Fuzz") Crompton, then the director of the South African Museum in Cape Town. (Fuzz is now the director of the Museum of Comparative Zoology at Harvard University; years ago he left his native South Africa, first to become director of the Peabody Museum at Yale before moving to Harvard. He is a big man and an outgoing person. We looked forward to our Stormberg trip with much anticipation; we expected it to be an interesting experience and we were not disappointed.)

On July 8 we left Cape Town in a British version of a jeeplike car, called a Gypsy, and drove over the folded mountains of the Cape through du Toits Cloof Pass, dropping down from there into the western edge of the Karroo. The scene changes rapidly. First there are the rugged slopes of the mountains, with spectacular euphorbias and proteas, these latter with large colorful flowers of which I never tired. Then into the Karroo with a progressively more arid biota, and with steep mesas and buttes the likes of which I had seen on my trip with James.

The night was spent in Victoria West, and on the morning of the following day, which was cold, we left for Bloemfontein. This day was made memorable by two events. One was the sight of an aardvark along the roadside. This was unprecedented: aardvarks are supposed to be strictly nocturnal but this one was breaking the rules of aardvarkdom. Fuzz jammed on the brakes, and I jumped out of the car hoping to get his picture, but he disappeared at high speed and went down a burrow before I could unlimber the camera. The other event was the discovery by Fuzz that a nut holding the right front wheel on the axle of the car had sheared, so there was nothing to keep the wheel from parting company with the vehicle at any moment. We had a slow, tense drive into Bloemfontein.

Much of our Stormberg trip was spent in the Herschel District of the Orange Free State in a southwesterly extension of the Drakensbergs, right up against the border of Basutoland (now known as Lesotho) and in Basutoland itself. Lesotho is a mountainous "island," surrounded by South Africa. During the last century a knowledgeable chieftain or king named Moshesh ruled the Basutos. He saw the Boer voortrekkers coming at him from the southwest and the Zulus descending upon him from the north. So he asked for help from the British government, and thus the mountain kingdom, a British protectorate, was established. Here the massive Drakensbergs rise to their greatest heights as a central range in the middle of the country, and from this uplifted mountain mass various rivers, cutting through deep gorges, make their tortuous, anastomosing journeys to the south and west eventually to join with the headwaters of the Orange River, which leaves Lesotho at its southwest corner to flow westwardly across the breadth of South Africa, reaching the South Atlantic Ocean at Alexander Bay some 350 miles north of Cape Town.

We journeyed to the Herschel District by entering the northwestern corner of Lesotho and traveling south through the western part of that country, finally to exit at its southwestern corner for a short drive to our destination. It was perhaps a roundabout way to get there but a paleontologically profitable route for us, and it allowed us to spend several days traveling through Basutoland.

Near Leribe, in northern Lesotho, we saw some unusual fossil footprints, enormous five-toed impressions made by a very large, heavy animal, probably one of the gigantic dicynodont reptiles and perhaps one of the last of its kind. Here we saw also various dinosaurian footprints which are the expected trackways to be found in Upper Triassic rocks. We then journeyed to the south through Basuto villages to the capital of the country, Maseru, and from there we continued south.

High on a mountain road we stopped at a place where we overlooked a Basuto village, a clump of picturesque round houses, known as *rondavals,* with conical thatched roofs, and nestled against the side of a mountain. While Margaret sketched the scene a group of Basuto boys gathered around us to watch the proceedings. Fuzz photographed them with a Polaroid camera and passed the pictures around, much to their delight. In return they sang for us—a part-song of unstudied charm.

A little farther on we encountered a lone musician, a stockily built man clad in very ragged trousers, a heavy blanket thrown carelessly over his bare torso, and a knitted woolen cap on his head. He was walking along playing his instrument, which consisted of a battered square tin receptacle punched full of holes (this was the sounding box) attached to an upright stick, a single taut wire extending diagonally from the top of the stick to the can. He bowed the wire with another stick, held in his right hand, controlling the pitch by pressing on the wire at various places with his left hand. Here was music from ancient days (the tin can might just as well have been a gourd or perhaps a turtle shell) that pleased this primitive man in a modern world.

Our next encounter on the high mountain road was with a group of Basuto horsemen, muffled to their eyes by their blankets, riding side by side on shaggy ponies. The Basutos, if not unique, are at least unusual among Africans for being horsemen. We stopped and they stopped and we had a little talk. They were on their way to a trial.

Near Morija, a little town south of Maseru, Fuzz and I climbed a long slope to look at some Red Bed exposures while Margaret visited with some Basuto housewives in a little cluster of dwellings at the foot of the hill. While we were clambering over the rocks looking for signs of fossils a group of about twenty men erupted from the settlement at the base of the hill and came yelling toward us, brandishing clubs and sticks. To use a well-worn phrase, my heart sank right into the toes of my boots, for I thought my last hour had come. I learned later that Fuzz had similar feelings. On they came, waving their clubs and shouting—and they went right past us as if we did not exist. It turned out that they were chasing a hare, which they failed to catch, having expended far more BTUs in the chase then ever they would have regained if they had caught their victim. But for one brief moment Fuzz and I thought that perhaps doomsday was upon us, as did Margaret, who saw the proceedings from below.

Finally, before leaving Lesotho for the Herschel District we visited a French Protestant Christian Mission, where its founder, the Reverend Ellenberger, had built his home by raising a brick wall beneath an overhanging ledge of Cave Sandstone, thereby enclosing a long and very substantial living area. It was a picturesque dwelling, the long brick wall punctuated by doors and windows and looming above it the massive cliff. It was made doubly interesting for us because on the ceiling of the living room was a trackway of dinosaurian footprints, evidently made by a little coelurosaur perhaps not unlike *Coelophysis* that we had collected some years previously at Ghost Ranch, New Mexico. And out in front of this cliff dwelling a large area had been enclosed within a tight fence, to preserve inviolate the native plants against the depredations of cattle. There we could walk within a bit of old Africa, pushing our way through grass that swayed head-high and higher.

Fuzz had made plans for our sojourn in the Herschel District; we stayed at Fort Hook with Mr. and Mrs. Malcolm Hepburn and their three little girls. The Hepburns ran a trading post at Fort Hook, and to this little commercial center, a sort of oasis surrounded by rather barren grasslands with interspersed cornfields, and always the dramatic hills and mountains in the background, came the varied black people to buy their supplies. All around us were the Tembus, farmers and shepherds, the men clad in orange-brown woolen blankets fashioned into capes, fastened at the right shoulder with a large safety-pin type of clasp. On his chest each man wore a dassie-skin tobacco pouch, suspended from his neck by thongs, and on his head a black hat. The men invariably carried stout sticks or knobkerries, and for our benefit they staged some mock stick fights that were almost terrifying to behold. The women were more colorfully dressed, usually in bright-colored blankets worn over beautifully decorated skirts and blouses. They wore turbans on their heads, and they were plentifully supplied with ornaments—necklaces, bracelets, and heavy masses of metal hoops around their ankles. Men and women smoked long-stemmed pipes.

From this interesting center we journeyed in different directions to look at the geology and to study fossil sites. By "we" I mean Fuzz and Rosalie Ewer and myself. (Margaret all this while was journeying around the countryside with an African guide and driver, making a collection of Tembu cultural objects for the American Museum and learning something about the people and how they lived.)

A word about Mrs. (or Dr.) Ewer, who had joined us at Fort Hook. To all of us she was invariably known as "Griff." Her husband was then professor of zoology at Rhodes University in Grahamstown, and Griff was a very accomplished zoologist and paleontologist in her own right. Griff was not cut from the usual piece of feminine cloth; she almost always wore trousers, she had a very short bob, and she smoked a pipe. Yet appearances can be deceiving, for Griff was a lovely woman, a blithe spirit in an all too somber world. When we knew her she had two grown children, and to compensate for their absence from home she had a pet meerkat (a sort of mongoose) which made life in the house more than interesting. Subsequent to our visit to South Africa the Ewers moved to Ghana, where they taught in the university. Then a few years ago Griff died of cancer in London. Quite characteristically, she entered a hospital, did not tell friends about it, and quietly lived out her last few days. She was one in a million.

We were at the Hepburn home, watching the black people come and go and having good times with them, and going ourselves on our different missions. For Margaret the days were extraordinarily interesting; among other things she attended some Tembu initiation parties and dances (she being the only white person present), and she had ample opportunities to visit with the women and learn something of their crafts, especially weaving and pottery making. For Fuzz and Griff and me the daily trips led us into the fastnesses of this mountainous region, where high on the cliffs we

The men invariably carried stout sticks or knobkerries, and for our benefit they staged some mock stick fights that were almost terrifying to behold.

found the remains of primitive Triassic dinosaurs in the Red Beds and the Cave Sandstones.

The time passed pleasantly. Every morning Margaret and I were awakened by the big Hepburn dog, who came into our room to say good morning. Every evening we had a good dinner with the family. One day, when we were not fossil hunting, we went on a picnic with the Hepburns during the course of which we visited a rock shelter with exquisite Bushmen paintings of elands, thousands of years in age. In technique and the perfection of the draftsmanship they were uncannily similar to the cave paintings made by Magdalenian men many thousands of years earlier in southern France.

Finally it was time to leave, and again, as was the case at the Brodie Farm, I reluctantly said good-bye to our hosts. We were off to the other side of Lesotho, getting there by circling around the southern border of the country. At Matatiele we drove westward, to cross, within a few miles, the eastern border of Basutoland, entering the country at a place with the picturesque name of Qachas Nek. Our destination was a mission known as Christ the King, in the very center of the country. There we stood at the edge of a terrifyingly deep canyon to look down upon the headwaters of the Orange River.

There is little more to say about the Stormberg trip. We returned to Cape Town along the southern coast of South Africa, a trip through a beautiful and historic land of old Dutch farms, bordered by the strand of the Indian Ocean.

We drove back to the west through East London, Grahamstown, and Port Elizabeth, where I saw in the new museum huge sauropod dinosaur bones from the Uitenhage Formation of Cretaceous age—my first sight of these fossils that I had previously encountered in the literature. At Mossel Bay, in the shallow water, I saw for the first time living tunicates, those saclike prevertebrates. As we drove on through the green countryside of the coast, I had time to think about Africa and what I had learned on my field trips. I had seen the sequence of Beaufort beds with their profusion of mammal-like reptiles, some of them closely related to the cynodonts and dicynodonts that I had collected in the Triassic sediments of Brazil, and I had seen the Stormberg Series, with Triassic dinosaurs related to those of Europe and North America. What did it mean? For one thing, it meant that Gondwanaland was becoming more real to me than hitherto it had been. There was no denying the close resemblances between the Triassic reptiles of South Africa, Brazil, and Argentina. For another thing it meant that there were other Triassic relationships as well, particularly toward the end of that geologic period, when similar early dinosaurs inhabited the northern and the southern continents. Pangaea, the immense ancient land mass that included Gondwanaland in the southern hemisphere and Laurasia in the northern hemisphere, also was becoming more real to me. The prophetic views of Wegener and of du Toit were taking on an ever-increasing appearance of actuality; they seemed less problematical than they had a year or two before.

After some more time in Cape Town we went back to Johannesburg, where my African experience had begun. There, one day, we visited Dr. Edna Plumstead, an outstanding paleobotanist at Witwatersrand University. I talked at length with her (she was a devoted "Drifter") and I saw a collection of fossil plants from Antarctica that she was studying. These plants, ranging in age from the Devonian into the Jurassic, were most impressive. I wrote in my notebook: "What does this mean with regard to past continental relationships? Could such plants have developed in the southern polar regions, even if temperatures were favorable, in view of the long polar night? This flora raises some very important questions." Indeed it did—questions with which I would become personally involved within the next few years. But I had no inkling of that at the time.

Well, the African visit was almost over. But not before we had a trip through Kruger Park with Ian Brink and his little daughter, Olga. There we had close encounters with elephants and hippos, lions, hyenas, jackals, zebras, giraffes, and innumerable antelopes of many species. For me it was like a journey back to the Pleistocene, when man in the Old World was surrounded by a host of large, aggressive mammals.

I was hoping to see something of Pleistocene man in Africa, for we had a date with Louis Leakey in Kenya. But just before we were ready to leave Johannesburg a telegram came from Dr. Leakey saying that he had been called away—could we come at a later time? We could not; our schedule could not be accommodated to such a contingency. We went to Kenya anyway, where some of Leakey's colleagues were very good to us. We got to see something of the great African rift, the northern limits of which we had already seen in Israel, but unfortunately there was no trip to Olduvai for us.

Then on to Greece, for a look at man in what seems to us as his Golden Age. After that to southern Germany, for the meetings of the Deutsches Palaeontologische Gesellschaft, and some very satisfying excursions through the fields and forests of Württemberg.

Finally, as a very special and fitting ending for this extended Triassic journey, we went to southern Switzerland, where Professor Emil Kuhn-Schnyder of Zürich was excavating Triassic fossils in a quarry on the very top of Monte San Giorgio, overlooking Lake Lugano. How could one have a better combination—fossils and a magnificent view across southern Switzerland? We stayed with Professor and Frau Kuhn-Schnyder and his students for a week or so in a commodious house in the little village of Merida, a mile or two from and a good many hundred feet below the fossil site. It was a wonderful vacation, paleontologically rewarding, with a good measure of sight-seeing thrown in on the side.

September was well advanced and Margaret had left, because she had to get back to the family. I still had things to do in England; people to see, meetings to attend, fossils to examine. These things I did and at last in mid-October I was back home.

13.
LAND OF THE GONDS

Our big Air India jet circled through the haze above Calcutta, and from the windows Margaret and I had recurrent glimpses of the Hooghly River and the massive Howrah bridge, first from one side of the plane and then from the other. At last we were down at Dum Dum Airport, where we were met by an old and a new friend—Dr. Pamela Robinson of the University of London and Dr. Bimalendu Raychudhuri, both emissaries from the Indian Statistical Institute in Calcutta. They draped garlands of flowers around our necks and then hustled us into a car for the drive to the institute.

Pamela was instrumental in arranging for our trip to India to participate in some of the work of the institute. The year was 1964, and we arrived in Calcutta on January 10. We were delighted to be with Pamela again; we had become acquainted with her previously when we were in London, and now here we were, in a strange land in the company of a cherished friend.

To our Western eyes the scenes along the way to the institute were exotic to say the least; indeed they were confusing and almost overwhelming. Hordes of people were everywhere, and our driver threaded his way among great crowds of pedestrians, wavering bicycles, rickshaws, creaking bullock carts, wandering cattle, and noisy trucks belching black fumes, all looming endlessly through a haze of smoke and a variety of wafted odors. Such was our introduction to India, the fabled land of Kipling, now brought more or less up to date.

In the matter of a half-hour or so we arrived at the institute, where we were again greeted, this time by new friends who soon were to become old and fast friends. Among those who welcomed us was Professor Prasanta Chandri Mahalanobis, the director of the institute, and several members of the institute staff: Sohan Lal Jain, Tapan Roy Chowdhury, and Pranab Mazumdar, all vertebrate paleontologists.

A few words of explanation are in order.

The Indian Statistical Institute was founded quite a number of years ago by Professor Mahalanobis, a statistician of worldwide acclaim. Early in its history Professor Mahalanobis decided that the institute should involve much more than statistics; he had a vision of a research organization in which there was an interaction between statistics and mathematics and various other scientific disciplines, including geology and biology. In line with this a department known as the Geological Studies Unit was established.

Pamela Robinson first went to India to help found the Geological Studies Unit. It was a formidable task, but Pamela is a dedicated person and she threw herself wholeheartedly into the project. In those first years her work was not easy; she must needs develop a program in field and laboratory—in the field often traveling by bullock cart and in the laboratory building a facility where hitherto there had been none. She succeeded.

Through the years she trained a number of Indian paleontologists in London, and she made numerous trips to India, where with the cooperation of her erstwhile students the policies and programs of the Geological Studies Unit were developed. Field studies and collections were made, especially in central India, in the Pranhita and Godavari valleys, while research based upon these studies and collections was pursued at the institute. When we were there Bimalendu was director of the unit, Sohan was working on fossil fishes, and Tapan on fossil amphibians and reptiles. Pranab was in charge of the technical laboratory. (Bimalendu died some years ago, still a young man, and Dr. Jain is now director of the unit. Tapan is still there and his researches are augmented by those of another paleontologist, T. S. Kutty.)

The institute is located perhaps ten miles away from downtown Calcutta on Barrackpore Trunk Road, which a century and more ago, in the days of the British Raj, was a suburban area wherein was located the residence of the Viceroy. Today the institute occupies a fairly extensive campus, with numerous buildings occupied by lecture rooms and laboratories, libraries, and residential quarters for some of the staff. Rows of palm trees, often bordering large ponds or "tanks," mark the institute grounds, and everywhere there are magnificent flowers. Overhead green parrots fly from tree to tree, and the air is often filled with the raucous cries of mynah birds. Around the grounds is a high wall, pierced by a number of entrances, each capable of being closed by heavy iron gates, each gate manned day and night, around the clock.

Outside the walls is the teeming life of Calcutta. On one side of the grounds heavy, noisy traffic roars along Barrackpore Trunk Road, while on

The Indian Statistical Institute occupies a fairly extensive campus, with numerous buildings containing lecture rooms and laboratories, libraries, and residential quarters for some of the staff. We had been assigned a pleasant apartment in one wing of Amrapali, *the headquarters building of the Institute.*

the other there is a lesser street, lined by shops and other establishments. The walls of the institute themselves form a sort of backdrop for many activities; peddlers' stands, little places to eat, parked trucks and bullock carts, all crowded between the walls and a roadside gutter that serves in part as a sort of open sewer. And along the roads, seemingly oblivious of the horrendous traffic, loose cattle wander, as they do in almost every Indian city and village. In fact, wandering cattle are to be encountered in the very heart of Calcutta.

Inside the walls there was for us an air of tranquility quite in contrast to the busy world just beyond their confines. In truth, the institute was a pleasant place in which to be, and that was a lucky thing, for immediately after we arrived terrible riots broke out in Calcutta, and a city-wide curfew was imposed. As the great metropolis came to a standstill we were restricted to the grounds of the institute, which was no real hardship. If we had been in a downtown hotel life would have been much harder to take.

We had been assigned a pleasant apartment consisting of a sitting room, a bedroom, and a bath, in one wing of "Amrapali," the headquarters building of the institute. There were several sets of living quarters in the building, one of them occupied by Professor Mahalanobis and his wife, Rani. (We did not get to see much of Rani on this trip, for she was very ill at the time and was confined to her bed, unable to see visitors except rarely and for short intervals. Fortunately she recovered and we became well acquainted with her subsequently, particularly on our second trip to India in 1977.)

Our days of involuntary confinement at the institute began with the coming of dawn and the noisy chorus of mynah birds. Then out from under the mosquito netting canopy that covered our bed to dress for breakfast, a protracted meal partaken at a long table set in a hall just outside the door to our apartment. Prasanta sat at the head of the table and we occupied chairs along the side together with Pamela Robinson and any other guests who might be staying at Amrapali. The meals were protracted because Prasanta was a great conversationalist, much more interested in discussing all sorts of problems than in eating. But he did get a bite down now and then, thanks to the insistence of his faithful servant, Bahadur, who stood by and almost forced Prasanta to eat.

Prasanta was truly an ascetic type of person, tall and very slender, his mind constantly occupied with scientific, economic, and sociological problems. He was an interesting, lively, and unworldly host, who stood high in the intellectual and political circles of India. He could not be bothered with the lesser, mundane matters of daily life, and to be blunt, he was accustomed to have people wait on him. We had this strongly impressed upon us one time, when we were all to fly to New Delhi together. As the time for the plane's departure approached I became anxious and uneasy, but Prasanta was not in the least disturbed. He continued talking to us as we sat in the reception room of Amrapali. Finally he decided it was time to go to the airport, so off we went in an institute car, with only minutes left in which to make the connection. I needn't have worried. When we arrived at the airport our car was motioned through a gate and we drove right up to the waiting plane. It was being held because Prasanta was booked for the flight, and it would have been held even if we had been much later than we were.

Prasanta was a close personal friend of Rabindranath Tagore, and some of Tagore's delicately executed pictures decorated the walls of the reception room and the dining area at Amrapali. He was also a friend of Nehru, who at that time was Prime Minister of India. Other friends and acquaintances included outstanding scientists and statesmen from around the world. Prasanta and Rani had no children, but they had a favorite niece, a stunningly beautiful, lively, and pert young lady, who, when normal conditions returned, came to Amrapali at times, to make our long sessions at the dining table very interesting indeed. She certainly was not cowed by Prasanta's distinguished position in the world.

Our meetings with Prasanta were mostly at the dinner table. During the day we visited with various staff members of the institute (likewise confined by the curfew) and had discussions about scientific problems, while trucks loaded with soldiers raced up and down Barrackpore Trunk Road. We never did learn the exact nature of the riots or their magnitude, but we did hear rumors to the effect that more than a thousand people were killed during the week of the curfew. At the end of each day, after the evening meal, Bahadur would come into our apartment and arrange the mosquito netting over the bed, and spray the place liberally with some sort of repellent, for even though there were screens on the windows the mosquitoes did get into the place—in large numbers. We would crawl under the netting, quickly tuck it in, and spend a few minutes demolishing any mosquitoes that might have entered with us. There was then a long tropical night before the mynah birds again would awaken us.

Finally there was a partial lifting of the curfew; people were to be allowed on the streets until five in the afternoon. We went into town that day, and the sight of thousands upon thousands of Indians hurrying home in midafternoon was something to remember. Most of the men wore the traditional white pajamas and punjabis, so the streets were filled with white waves of humanity—in, on top of, and hanging from the sides of buses, in cars, in rickshaws, on bicycles, and on foot. There was no doubt as to the desire of all the people to be off the streets by the time the appointed hour arrived. The penalties for being abroad after the curfew were severe. Quickly the curfew was relaxed and in the course of a few days life was back to the noisy, frenetic norm of Calcutta. It was then that we made plans for our first excursion into the field. Our destination was the Raniganj coal field, about a hundred miles to the northwest of Calcutta, where we would find exposures of the Lower Triassic Panchet Formation.

We took the train from Calcutta to Asansol—Pamela, Margaret, and I. At Asansol the jammed station platforms were kaleidoscopic, made so by the varied garments worn by the people, particularly the brightly colored saris in which the women were dressed. We pushed our way through the crowds to the waiting room, where the first thing we saw were some exquisitely executed figurines, made in clay and skillfully colored, of the Indian goddess of learning, Sarasvati. Her festival was approaching, a time when shrines are set up all over West Bengal in her honor, containing countless numbers of statues such as the ones we saw. We had to stop to admire them.

We were met by Bimalendu, who was accompanied by Dr. P. P. Satsangi of the Geological Survey of India, they having driven up from Calcutta ahead of the train. A jeep was waiting; we piled in and were driven through the crowded streets of Asansol, past creaking bullock carts and bicycles and pedestrians, and then we were out in the country on the way to our camp near the little village of Tiluri, to the south of the Damodar River.

Our camp was more than a camp, because Bimalendu or somebody in authority had obtained permission for us to bivouac at a country health center then under construction, but at which work had been temporarily suspended. Thus we had some solid buildings in which to sleep and eat; Margaret and I set up our camp beds in a neat little edifice destined to be quarters for a nurse or a doctor, Pamela did the same in another building, Bimalendu and Satsangi in still another. Our support personnel, consisting of Ram, a huge, solid man with a fierce moustache, Debraj the cook, and a couple of helpers, stayed in a large unfinished building that was to be the hospital. There Debraj established a kitchen, and nearby Ram arranged tables at which we all could eat. A mile or two away from our quarters was Susunga Hill, a small mountain rising from the plain as a prominent landmark in this region, while in another direction we could see the temple in Tiluri, a picturesque sight when viewed through the early morning haze rising from numerous cooking fires.

Each morning Ram would appear with tea, which was consumed before we crawled out of our beds. After a proper interval we would walk over to our dining room for breakfast, following which we would go our several ways.

Our several ways involved one direction for Margaret and another for the rest of us. Arrangements had been made for Margaret to see something of the Santal people, an aboriginal tribe that lived in this area, and she was provided with a jeep and a driver and an Indian docent.

The Santals live in their own villages, quite apart from the Indians, and their villages are quite unlike Indian villages. For one thing, they are tidy and immaculate; the streets and paths are swept free of litter and trash and the houses are neatly whitewashed. This impressed Margaret's Indian guide, who expressed some amount of wonder concerning the well-ordered Santal way of life. They have their own cultural patterns, including rituals and dances. Margaret attended a ritual which involved a chorus of maidens, neatly clad in saris with marigolds in their hair, some dancing, a weird performance by a man who must have been drugged, and finally the slaughter of a goat by decapitation, following which the participants in the ceremony rushed to drink the blood as it spouted from the neck of the headless animal. She found this to be pretty strong stuff. Nevertheless she found the Santals to be very interesting and friendly people, and she treasured the days spent with them.

For the rest of us our path each morning led into the rock exposures, where we studied the geology and looked for fossils. We would hike along, scrutinizing the outcrops and discussing the possible significance of what we were looking at while Ram brought up the rear with a haversack on his back. In that sack were our geologic hammers, and when we wished to use our hammers we would go over to Ram, take them out of the bag, still on his back, and whack away at the rocks. When we were through with our hammering we would go back to Ram and return the hammers to the bag. We quite obviously were not supposed to carry our hammers as is the usual custom among geologists and paleontologists, for to do so would have been to deprive Ram of a task that clearly was his.

Division of labor is an important aspect of life in India. For example, none of us were to drive a jeep. Each jeep had its driver, whose task it was to drive and to keep the car in good running order. When the car stopped and we left it to explore the terrain on foot, the driver remained with the vehicle. A story. A young Englishman, Peter Johnson, was doing geological fieldwork under the auspices of the institute when we were there. One day he had his driver stop the jeep, so that he could walk down a nullah for a bit of geologizing. Peter vanished into the jungle and a minute or two later a large tiger appeared and followed him down the nullah. The driver, sitting in the open car, was petrified with fear and dared not move a muscle. He was in a fever of excitement and indecision as to what he should do. It was a frightening situation, but before he could make up his mind on a course of action Peter reappeared, having made a circle through the jungle quite unaware that he was being followed. Evidently the tiger was merely curious; at any rate Peter and the driver did not try to look him up for questioning.

To get back to our story, on our third day in the field I had the pleasure and the thrill of finding a *Lystrosaurus* skeleton, already well known in the Panchet Formation. (It may be remembered that James Kitching and I had collected *Lystrosaurus* in South Africa.) There was for me this day an emotional as well as a scientific experience in seeing this reptile in the field, *in situ,* at a locality now thousands of miles removed from its African counterparts. It was incontestable evidence for a dry land connection between India and Africa in early Triassic time. Why could not the connection have been a short one, the result of the former presence of India within the continent of Gondwanaland, the western border of India contiguous with a part of the eastern border of Africa? This *Lystrosaurus* near Tiluri might not prove the one-time drift of India northward from its former connection with Africa, but it certainly did point to such a possibility.

We poked around the fossil (what I had first seen was the side of the skull, with the orbit exposed) and after some time it became apparent that the entire skeleton might be there in place. We made plans.

The plans involved going back to our camp, assembling all of the proper equipment and materials, and coming back to dig out the skeleton. On the following morning we turned up in full force at the *Lystrosaurus* site, armed with picks and shovels, awls and scrapers, shellac, brushes, sacking, and plaster of Paris. Then we started our task in the usual way.

Excavation of the *Lystrosaurus* skeleton required two days of our time, with a considerable audience looking on. People are everywhere in India, and we were hardly well started with our work on the first day, when some small boys appeared, soon followed by some of their elders. Before long there was a group of a dozen or so men and boys squatted on the side of a little hill a few feet from where we were working, watching our efforts with the greatest of interest. The audience kept changing as people kept coming and going, but it was there all of the time. Interestingly, it was an all-male audience; I suppose the women and girls were too busy with their usual duties for such idle pursuits.

At last on the afternoon of the second day the fossil was jacketed and ready to be moved. It was by now a fairly heavy specimen, so a sling was made for it out of strong sacking, a long pole was thrust through the sling, three of our party took hold of one end of the pole and Ram, a mighty man, took the other end. In this manner the fossil was carried to the jeep waiting at the top of the hill, with three idle water buffalo reluctantly giving way to our procession as it went on its way .

Thus ended my first field experience in India—in a highly satisfactory fashion. I had found a nice *Lystrosaurus* skeleton, which in itself was valuable. Beyond that were the implications of the find, particularly cogent in

Excavation of the Lystrosaurus *skeleton required two days of our time, with a considerable audience looking on. P. P. Satsangi (far left) and Bimalendu Raychudhuri (on the right in light-colored shirt).*

A sling was made for the fossil out of strong sacking, a long pole was thrust through the sling, three of our party took hold of one end of the pole, and Ram, a mighty man, took the other end. In this manner the load was carried to the jeep waiting at the top of the hill.

my mind because of my personal involvement, not only with *Lystrosaurus* in India, but in Africa as well. Gondwanaland was becoming ever more meaningful to me.

Now came an interval of traveling and sightseeing. We returned from our field trip to Calcutta, and then flew to New Delhi for the Indian Republic Day parade and celebrations. The parade was impressive. A colorful and seemingly endless procession marched before us down the broad Rajpath, which with its parklike borders forms a triumphal way through New Delhi. We were sitting in one of many grandstands that had been erected along the Rajpath, fully exposed to bitter cold winds that seemed to come right down on us from the Himalayas, and as we sat and shivered, we watched, with a fascination that made us almost forget the frigid breezes, the magnificent spectacle that passed in front of us. There were numerous and varied army units in smart uniforms, including the famous Gurkhas, noted for their fighting abilities. There were Sikhs on foot, and mahouts on richly caparisoned elephants, and there were camels. There were colorful floats. At the end of the affair thousands of rubber balloons and doves (dyed all pink) were released into the air, while fighting jets roared overhead, trailing long streamers of colored smoke. All of this took place in the morning.

In the afternoon we attended a garden party at the Presidential Palace. Since the day was more than cool I wore a little pill-box-shaped hat made of fake fur, to keep my bald pate warm. During the course of the afternoon Lord Mountbatten, who had been the last Viceroy of India, and now resplendent in a naval uniform, circulated through the crowd on the lawn, accompanied by a coterie of Indian dignitaries. For an instant we were face to face, and during that instant Lord Mountbatten recoiled ever so slightly with a somewhat startled look. As he saw me in my head-warmer perhaps the words of the Prince of Morocco in Shakespeare's *Merchant of Venice* flashed through his mind—"O hell! what have we here?"

The garden party over, we had a trip to Sikandra, and from there to Agra, to see the Taj Mahal. Margaret and I had read so much about the Taj and had seen so many pictures of it that we were afraid we might be disappointed when we actually encountered it. We were not. After that to the old, long-abandoned city of Fatehpur Sikri, built by Akbar, and finally back to Agra and to Delhi.

Pamela, Margaret, and I boarded a plane in New Delhi on the evening of January 28 for a flight to the city of Nagpur in central India, which was to be the starting point for our next foray into the field. We shared the plane with a group of Russian technical advisors and their families—the women plainly dressed, the men wearing suits that somehow seemed too baggy. When we arrived in Nagpur it turned out that all of us were bound for the same hotel, a modest establishment that turned out to be far from luxurious. We all arrived at the hotel at the same time, late at night, and there was some confusion before we got properly settled in our respective rooms, which were a bit spartan.

Our party was to consist, in addition to Pamela, Margaret, and myself, of Dr. Jain, Tapan Roy Chowdhury, Pranab, Debraj, and two drivers for the jeeps. Each of the jeeps pulled a trailer, so we had the means to haul all of our luggage and equipment to the places where we would be working. The next day after our arrival in Nagpur, where we met the other members of the party, we devoted ourselves to getting ready for the trip. However, we had some time to see the city and its environs, and I shall never forget our excursion into the middle of town, where all wheeled traffic, except for bicycles and rickshaws, was prohibited. The scene was wonderfully quiet; it was truly a relaxing experience to walk through the streets, hearing only the footsteps and the chattering voices of the pedestrians. What a blessing it would be, I thought, if all cities could be like this—at least in their central cores.

Our first day of driving took us to Chanda, an ancient walled city now only partially occupied. The night was spent at a government rest house, such as are found throughout India. These rest houses were established many years ago by the British rulers of India as conveniences for people traveling on official business. They are still maintained. One goes to a rest house, bringing one's own needs—food, bedding, and the like. Then if the rest house is available one can for a small fee, as I understand the matter, bed down and prepare meals. The matter of availability is decided by protocol, and each rest house has posted in it a list of priorities. Of course the Prime Minister heads the list (if he should ever deign to stop at a rest house) and after that comes a descending order of eligible users, down through high government officials to judges and so on, with geologists and biologists at the very bottom of the page.

On the next morning we went in a jeep over country roads to the confluence of the Wardha and Penganga rivers, where we took a boat, a sort of dugout, for a trip of about three kilometers up the Penganga to a place near a little village called Irai, where a glaciated pavement was to be seen. Here Precambrian rocks are beveled by a smooth surface marked with glacial striae, grooves plowed into rocky floors by the grinding movements of glaciers carrying their abrasive loads of rocks and pebbles. The morphology of these striations shows that the glacier was moving toward the northeast. Just above the glaciated surface is the Talchir boulder bed, an accumulation of heavy, unsorted boulders, evidently deposited by glacial waters, while above the boulder bed is the Talchir Formation, a khaki-colored sandstone that can be dated as of Carboniferous age. This is geologically a particularly significant locality, for at first glance it would seem to be an isolated northern hemisphere record of Paleozoic glaciation, far removed from correlative indications of ancient glaciation to be found in southern Africa, Brazil, and Australia. But if India were at that time a part of Gondwanaland, then this glacial evidence makes sense. It is part of a larger record of a great glacier that had its center in southern Africa, the glacial lobes pushing out in all directions, with those covering what is now India pushing in a northeasterly direction. In accordance with the theory of continental drift, India subsequently drifted to the north, to its eventual collision with the great Asiatic land mass (pushing up the Himalayas in the process) and carrying the evidence of its ancient glacial history far away from where the events originally took place.

This was another bit of evidence for Gondwanaland and the drift of India, evidence to be added to that of the *Lystrosaurus* skeleton we had dug up only a few days earlier. As we walked around over the smooth glaciated rocks a herdsman brought his cattle across the ancient surface to drink at the river, while a hundred yards downstream, at the edge of the water, a funeral pyre was blazing, the half-consumed corpse visible through the hot flames.

Our journey continued to Sironcha, a large and commodious rest

Here Precambrian rocks are beveled by a smooth surface marked with glacial striae, and the morphology of these striations shows that the glacier was moving toward the northeast.

house, well sited with a fine view of the Pranhita River. On our way we drove through the teakwood forest, not at all like my mental picture of an Indian jungle, for it resembled to some degree a hardwood forest in the Middle West. Yet we knew that this was the habitat for many animals, sambar deer and foxes, leopards, and now and then a tiger. As we came to a river crossing in the late afternoon we surprised two Gond men, who viewed us with some suspicion. As we stopped the jeeps and began to dismount they left in a hurry, disappearing into the forest before we had a chance to speak with them.

Gondwanaland was named by Eduard Suess, a famous Austrian geologist, late in the nineteenth century. The name is derived from the land of the Gonds, once inhabited by a Dravidian people who a millennium ago had established a kingdom in the central part of India. Today the Gonds remain as a remnant of the former population, living as aboriginal bands deep in the forest and isolating themselves as much as possible from the Indians who surround them. Suess coined the name Gondwanaland to indicate a former hypothetical continent embracing central India, Madagascar, and southern Africa, because he saw similarities between the rocks in these regions. (Subsequently South America and Australia were included in Gondwanaland, and eventually Antarctica as well.) That was long before Wegener's original concept of continental drift. Suess envisaged Gondwanaland as a vast supercontinent with broad land masses bridging the oceans between the southern continents in their present positions. He thought, as did many geologists for decades afterward, that large portions of Gondwanaland had foundered into the oceanic depths, the remnants of the ancient supercontinent persisting today as the southern hemisphere continents plus India. Theories of plate tectonics and continental drift have greatly altered our views as to the original configuration of Gondwanaland and the manner in which the continents derived from it have reached their present positions. But Gondwanaland in its modern version remains as a very integral part of the earth of long ago, that part for which it was named being the land of the Gonds. We were in the land of the Gonds.

At the Sironcha Rest House we made ourselves comfortable and settled down to stay for several days. There was a large tree right next to the building, just outside the room where Margaret and I slept. During the day it was inhabited by a gaggle of flying foxes, the huge, fruit-eating bats of the Orient. They were anything but quiet during their diurnal hours of supposed rest; it seemed as if they were squabbling all day about one thing or another. At dusk they would flap out into the air for their nightly foraging, and to see them against the waning light, their large leathery wings (a

The teakwood forest was not at all like my mental picture of an Indian jungle, for it resembled to some degree a hardwood forest in the Middle West. Left to right: S. L. Jain, Pranab Mazamdur, Margaret Colbert, Pamela Robinson, Tapan Roy Chowdhury, two assistants.

yard and more from tip to tip) beating slowly up and down, reminded us of the days of the dinosaurs, when the aerial reptiles known as pterosaurs flew through Mesozoic skies.

The institute had a considerable quarry near Sironcha, opened in 1961, where under the direction of Dr. Jain a rich deposit of large dinosaur bones was being excavated. These were the bones of a sauropod dinosaur, the sauropods being the giants among giants, and these fossils at Sironcha were particularly significant because they were being found in rocks of basal Jurassic age. They were the oldest sauropod bones known, demonstrating quite clearly that the sauropod dinosaurs became giants at the very beginning of their long evolutionary history. (More than a decade later the bones, having been cleaned and prepared, and christened *Barapasaurus* by Dr. Jain, T. S. Kutty, Tapan, and S. Chatterjee, were assembled into an impressive skeleton, fifty feet or more in length, now to be seen at the institute in Calcutta. It is the only mounted dinosaur skeleton in India.) Our purpose at Sironcha was to spend some time at the quarry with Dr. Jain and his assistants.

That we did, working beneath the shade of a brushwood shelter that had been erected to protect us from the Indian sun. The work was essentially similar to the quarry work that I have already described. Perhaps the

Dr. Jain (left) and Pranab Mazamdur and a dinosaur femur. The Institute had a considerable quarry near Sironcha, where, under the direction of Dr. Jain, a rich deposit of large dinosaur bones was being excavated.

one big difference was that here there were several assistants whose task it was to carry away the surplus rock and dirt resulting from the digging (the "tailings" in mining terms) and dump it at a little distance from the quarry. This was accomplished in the traditional Indian fashion by carrying the dirt in flat baskets, balanced on the heads of the carriers. In this matter of carrying individual loads perhaps the Indians are smarter than are people of the Western World. Instead of using wheelbarrows and putting leverage on the lower back, which goodness knows is all too subject to strains, they place the weight on the head to allow the force of gravity to carry down straight through the body and the legs. Carrying the tailings away from the quarry was necessitated by the fact that the dig was on flat ground, not on a hillside as so often is the case, thus allowing surplus material to be shoveled down the slope. In fact, the big bones originally had been found scattered over the surface of the ground like so many boulders.

As a result of our labors we managed to remove all of the significant bones that had been exposed in this particular quarry. On the final day the specimens, encased in plaster, were manhandled into the jeeps and hauled to the Sironcha Rest House for eventual transportation to Calcutta.

Some sort of celebration seemed to be in order. One of our Indian assistants had gotten in touch with the chief of a Gond village, about fourteen miles distant in the jungle, where that evening we had planned to go to see a Gond dance. But that evening we were all too tired from our struggles with fossils during the day to attempt the trip. Another emissary was sent to the Gonds and he thought that an agreement had been made for them to come and dance for us at Sironcha. At bedtime there were no Gonds, and so we went to our cots much disappointed.

The next morning while we were eating breakfast they appeared. There were about two dozen of them, men and women, and they had walked through the jungle during the night, two of the men carrying an immense drum, the drumheads being made of nilgai hide. (The nilgai is the blue buck of India, *Boselaphus tragocamelus,* a large antelope.) These Gonds formed an exotic and colorful group of people.

They were unlike the Indians in various ways. It seemed to me that their skin color was perhaps of a more coppery hue than is the case among the Indians. Certainly the women had rather frizzy hair. We couldn't see the men's hair because they wore large white turbans, some with peacock feathers thrust into the folds of the headdress. The men were clad in loose blouses and long kilts or skirts that were gathered between the legs to form baggy dhotis, similar to those worn by many Indian men. They were barefooted, as were the women. The women wore bands around their heads, from which were suspended on each side long, woven strands of cloth, colored red, white, and black. The women were plentifully supplied with beads, and each wore a small, rectangular mirror that hung down on the front of the chest. Each woman had a silver ring through the left nostril, and some had tattooed cheeks. They wore small shawls fastened at the throat, and very short wrap-around skirts—something no Indian woman would wear. The men and women alternately carried long wands, from the ends of which dangled chains like the paper chains we used to make when we were children.

Many of the men carried small drums (in addition to the giant drum mentioned above) and they also had some wind instruments, especially a sort of short horn made of metal and supplied with keys. How I wished a tape-recorder were available.

Incongruously, the chief or headman carried a large ordinary black umbrella during the entire proceedings.

The dance began, and soon there was a large crowd of Indian spectators in addition to our party. The Gonds and their dance obviously were about as much of a novelty to the Indians as to ourselves. I cannot begin to describe the intricacies of the dance, or the haunting sound of the music, which went on and on, for most of the morning. The two men who carried the giant drum, suspended from a long yoke over their shoulders, swayed back and forth as they beat out the basic rhythm, and a large banner that was attached to poles rising from the yoke between them also swayed, to emphasize the tempo of the dance.

It was a festive introduction to the center of Gondwanaland. I wonder if any of these Gonds had ever heard of Gondwanaland?

After leaving Sironcha we had to cross the Pranhita River.

We finally had to depart, but the Gonds did not wish to stop their dance. As we slowly drove away from the rest house in our jeeps they surrounded and escorted us for quite a long distance down the road, beating their drums and playing their instruments. It was, perhaps one might say, a festive introduction to the center of Gondwanaland. I wonder if any of those Gonds had ever heard of Gondwanaland?

After we left Sironcha we had to cross the Pranhita River, a considerable watercourse with no bridge at the place where we encountered it. There was a ferry the likes of which I have never seen before or since. It consisted of two canoelike wooden boats, each about sixteen feet or so in length and each with perhaps a four-foot beam, fastened together side by side and at a little distance each from the other. There was a platform between them covered with a bamboo matting. This catamaran-type of contrivance was pulled up parallel to shore, a pair of none-too-heavy planks was placed from the gunwale of the near-shore boat to dry land, and then it was up to the driver of the jeep to ease his vehicle onto the ferry platform without going over the other side into the water. It was tricky business, because the platform was barely wide enough to accommodate the short wheel base of the jeep, the planks bent alarmingly as they felt the weight of the car, and the ferry, likewise feeling the weight, listed heavily toward the shore. Our drivers were skillful and the loading was managed successfully. Four crossings were necessary, one for each jeep and one for each trailer, the trailers being pushed by hand onto the ferry. The crossing was accomplished by means of oars and poles, and of course the whole process was reversed at the opposite bank of the river. I was relieved when the last trip, on which I was a passenger, was finally concluded—so relieved in fact that I promptly slipped off the plank while debarking and plunged waist deep into the not very clean river water.

Our objective was a little village called Bhimaram, near which were Upper Triassic outcrops that we wished to look at. Bhimaram is located in the middle of what on a mid-Victorian map of India was marked as "unexplored jungle." It has since been explored and settled to a considerable extent, but there is still jungle with jungle inhabitants. While we were working around Bhimaram we saw leopard tracks on several occasions, and we knew that there were tigers in the vicinity.

Our party set up camp at the edge of Bhimaram, but Pamela, Margaret, and I had the privilege of staying with Mr. Raja Reddy and his family. It was, I think, Raja Reddy's father who came into the jungle here and established the village of Bhimaram, and became the headman. Raja Reddy, at the time of our visit a young man in his early thirties, a very dark-skinned, handsome Telegu, was carrying on the tradition and was the headman of the village. He and his wife were graduates of Osmania University in Hyderabad. They had three small children, and a comfortable house in which, among other things, there was an extensive library. Raja Reddy was well read and very much interested in the world around him. He was active in trying to improve the lot of the villagers; some of his efforts were in the direction of introducing and encouraging the development of improved crops of various kinds. When we were there he was much involved with hybrid corn.

Raja Reddy, being the sort of person he was, took an active interest in our fossil work. He was frequently out in the field with us and lent valuable support to our efforts. Our efforts were concentrated on the Triassic beds here—the Lower Triassic Yerapalli Formation, in which we found reptiles of definite African affinities, and the Upper Triassic Maleri Formation, in which there are big metoposaurian amphibians and phytosaurian reptiles that duplicate almost identical fossils found in central Germany and in North America, particularly in the Chinle Formation of the Southwest. It was evident that in late Triassic times Gondwanaland was being invaded by animals from the north—from ancient Laurasia. Yet the Gondwanaland affinities are there, because in the Maleri Formation are rhynchosaurs, those peculiar reptiles, examples of which I, together with Llew Price, Carlos de Paula Couto, and Fausto Luiz de Souza Cunha, collected in the Santa Maria Formation of Brazil. So we spent interesting but hot days in the countryside around Bhimaram, while I was becoming ever more familiarly acquainted with Gondwanaland.

One evening after dinner, as we were all visiting and enjoying the cool air after a hot day in the field, Raja Reddy suddenly proposed that we go out to look for tigers—or at least a tiger. It seemed like a harebrained scheme but it promised some excitement, and Raja actually had encountered a tiger one evening not long before our visit while he was driving at night through the forest. So we all piled into his Land Rover—Pamela, Margaret and I, Jain, Tapan, and a geologist who was working with our party, Supriya Sen Gupta. We barged around through the woods in the dark, in the process scaring up a sambar deer and a fox, but we never saw a tiger. I suppose any self-respecting tiger would have kept his own counsel and watched us rattle by. In the course of things we got lost, and I can still hear Sen Gupta shouting, "Find the forest boundary—then we will know where we are!" But how were we to find the forest boundary? At last we wound up in a dry rice paddy, and had the devil's own time getting the car up and over the dike that surrounded it. It was quite an evening, and we finally got to bed about one in the morning. Fossil hunting isn't always fossils.

However, we collected some good fossils from the Triassic beds around Bhimaram, and then Margaret and I set out on a roundabout trip, arranged by Professor Mahalanobis, to see some of the sights of India. We went to Bombay, and to Aurangabad to see the stupendous Ellora temples, and after that to Ajanta to see the equally stupendous "caves," carved in the hard, volcanic rock of the Deccan lava field. Then back to Bombay and from there to Hyderabad, where we were met by Raja Reddy and his relative, Justice P. Jagan Mohan Reddy. We were to stay with Justice Reddy and his family.

Justice Reddy at that time was Chief Justice of the High Court of Hyderabad; subsequently he was appointed to the Supreme Court of India. He was a big, burly man, who had been educated in England. Mrs. Reddy was a handsome woman, and their grown children were likewise most presentable. (All of the Reddy clan were good-looking people.) They took us in and we were immediately a part of the family—as we had been with the Raja Reddy family. Our stay in Hyderabad was interesting and delightful, particularly the meals at the Reddy table. Justice Reddy may have been a very important man, but it amused me to see the manner in which his children argued things out with him. They were not in the least cowed by his position and erudition. Thus our meals were times of lively discussion, gesticulations, and decisions. Through all of this Mrs. Reddy quietly held up her end of the debates; she was not to be put down by anybody.

We had other interesting trips in India. I won't try to describe them except to say that we went to Madras, where we stayed with Mrs. Swaminathan, a beautiful dark lady who always wore a snow-white sari—she was a friend of the Reddys (her husband was away at the time); to Benares, to Darjeeling, and from there to Takdah, not far away, where we visited Mr. Sen, one of the officers of the institute, and his cousin, Miss Sen, at her home. Perhaps I should conclude this tale of India by saying something about our visit to New Delhi, where we had tea with Prime Minister Jawaharlal Nehru and his daughter, Indira Gandhi—also the Prime Minister in later years. Nehru had requested the visit—evidently he was interested in what we had been doing. Professor Mahalanobis thought that we should tell Nehru about our work; he thought it might be beneficial for the Prime Minister to get his mind diverted from affairs of state for a few minutes.

We had various things to do in New Delhi, but on the afternoon of March 16 we got ourselves ready to be driven to the residence of the Prime Minister. Our party consisted of Professor Mahalanobis, Pamela, Margaret, and myself. Once we had arrived we were led by some sort of majordomo through long corridors and past guards to a sitting room, where presently Nehru and Mrs. Gandhi entered. He looked very tired. It was the time when China had invaded Tibet, and had placed her hordes hard against the Himalayan barrier, ready, it would seem, to pour down into the valleys of India. The Prime Minister had reason to be worried.

However, he did seem to throw off his cares for a while, and we had tea and cakes and conversation together. We told him about what we had been doing, and during the course of the visit I managed to beat the drum for the betterment of natural history museums in India. This seemed to interest Nehru particularly. And of course we talked about Gondwanaland. Also, Pamela brought up the subject of how difficult it was to get maps of India in India because the government had prohibited the sale of all maps—presumably so they would not fall into the hands of the Chinese. Yet Indian survey maps were readily obtainable in London. Indira Gandhi frowned at this; perhaps she felt that her father should not be bothered by such matters. But Nehru smiled and implied that bureaucrats would have their way. Finally our time was up, so we departed. And that's the way it was.

We returned to Calcutta and later that month we got ready to leave India for Australia, another segment of Gondwanaland that I was anxious to see. We did leave India with a certain feeling of reluctance. It had been a marvelous experience for both of us, and we had made many friends. It was hard to leave them.

And it was hard to leave this land of color and of contrasts. We had seen things that the tourists see, but we had poked into parts of India that are unknown to the usual travelers. We had seen the past splendors of this land and we had seen some of the modern problems with which it is beset. We had seen the unbelievable poverty and squalor of Calcutta, but we had seen the condition of the country people, who in spite of their hard lives can live with dignity. We had lived with Indians and worked with them, so we did not feel like strangers. As for me, my paleontological education had been measurably advanced by this visit to the one part of Gondwanaland now entirely in the northern hemisphere.

We left India hoping to return, and we did return thirteen years later. It was a happy return, to old, familiar places and to old, firm friends. Unhappily some of our friends were gone—notably Professor Mahalanobis and Bimalendu. But the others were there, and we acquired some new friends as well.

When we departed from India that first time we had no way of knowing that we would be back. So we said farewell with a feeling of sadness at having to leave the subcontinent—the land of the Gonds.

14. DOWN UNDER

We flew from India to Australia with a stop in Thailand to see some sights. In so doing we journeyed from the Orient back to the Occident, from the complex and colorful cultures of the subcontinent to the rather prosaic Western World of the Island Continent. We felt the change in many ways, one of which was our craving for meat, which we satisfied during our first week in Australia by having steaks for breakfast as well as at other times. Soon we became adjusted once again to the Western way of life.

To the evolutionist Australia is a land of particular fascination, for here is a continent that has been isolated from the rest of the world during perhaps 50 million years—from Eocene time to the present. The isolation began before placental mammals had gained a foothold in this fragment of Gondwanaland; consequently, Australia is the land of marsupials or pouched mammals, and of the monotremes, those very primitive mammals, the platypus and the echidna, which reproduce by laying eggs. It is a land whose original human inhabitants, the Aborigines, had a remarkably primitive and different culture unique to this isolated continent. For the student of recent life Australia affords a visual projection into earlier ages when plants and animals and men were set apart in many aspects from the plants and animals and men in the rest of the world.

Australia offered me an opportunity for an even longer view into the past, to a time when what is now the Island Continent was still an integral part of an unbroken Gondwanaland, to a time when life here was closely related to life throughout the extent of the ancient supercontinent. To gain this distant view I hoped to give particular attention to three Australian regions—to the eastern coast, especially areas in New South Wales, Victoria, and Queensland; to Tasmania; and to the lonely northwestern coast in the vicinity of the little towns of Derby and Broome.

It may be recalled that in central India we took a little boat up the Penganga River to see a glaciated rock pavement, evidence of a time long ago when India was attached to Africa and was partially covered by a great continental glacier pushing out in all directions from a South African center. Now that we were in Australia it was our good fortune to journey one day with colleagues from Melbourne to a place known as Bacchus Marsh, and from there to a locality where there was a glaciated pavement. Here, once more, was evidence of that ancient age when Gondwanaland was the site of extensive glaciation. Australia at that time probably formed the easternmost extension of the supercontinent, and the immense ice sheet, advancing to the east from its South African center, scoured the rocks of what is now southeastern Australia. The movement naturally was in an easterly direction, but owing to the subsequent migration and rotation of Australia, the scratches on the rocks near Melbourne are today in a northerly trend.

My primary interest was of course in the fossils of Gondwanaland, so with the aid of Australian friends, notably Edmund Gill of the Victoria Museum, Jim Warren of Monash University, Harold Fletcher of the Australian Museum in Sydney, and Jack Woods and Alan Bartholomai of the Queensland Museum, I was able to poke around in Triassic sediments where fossils had been found. I didn't find anything in particular, but at least I became familiar with the setting. While in Queensland Margaret and I took advantage of the opportunity to visit Heron Island on the Great Barrier Reef, where we saw the colorful marine life of the reef, and where we watched little sea turtle hatchlings come out of their eggs, buried in the sand, to scramble in frantic haste from the beach toward the rolling breakers of the South Pacific Ocean. It was on the Barrier Reef that Margaret and a very nice young marine biologist from Kuala Lumpur, Miss Goh, were followed by a shark while they were snorkeling. They reached shallow water and stood up facing their pursuer, ready to sell their lives dearly, but at the last moment the shark, having second thoughts, made a 180-degree turn and swam out to sea.

I was especially anxious to see the Triassic beds of Tasmania, where Jim Warren had worked, and where at the time John Cosgriff, now at Wayne State University in Michigan but then on a fellowship, was working. Margaret and I took the plane to Hobart, where we were met by John and his wife and two young sons. Then there began for us an almost idyllic time in southern Tasmania. It was April—full autumn in that part of the world—and we enjoyed many a fine day in the Derwent Valley north of Hobart and on the beaches around the city where Triassic rock outcrops were to be seen. When we were not out in the field we were in Hobart, a small but very attractive city that captured our hearts.

It was there that the newspaper one day carried a story about a huge

Russian whaler anchored in Storm Bay, about thirty miles south of Hobart. The paper sent a light plane out to circle the whaler and as a result published what was their interpretation of its name. Margaret, who had been studying Russian, was able to send a note to the paper with a correct transliteration of the name of the ship; it was the *Sovietskaya Ukraina,* obviously waiting for instructions from Moscow.

Shortly thereafter we flew back to Melbourne and at the same time the Russian vessel put in there, where open house was held for interested visitors. The ship was almost overwhelmed by hordes of Melbournians. For several days in Melbourne we saw little knots of Russian sailors walking the streets and entering the stores, always close to each other and evidently somewhat ill at ease in a strange land.

At the end of our travels and studies in the eastern part of Australia Margaret left to return home, while I stayed on for a western Australia adventure.

Perth is one of the most isolated cities in the world—to the east some 1,500 miles from Adelaide and more than 2,000 miles from Sydney, 1,500 miles from Java to the north, 4,000 miles across open ocean from Madagascar and Africa to the west, and another 4,000 miles due south to the South Pole. In former years Perth felt its isolation; the trip from Adelaide by train across the great Nullarbor Plain was time-consuming and tedious. Today things have changed, and Perth is very much a part of the world, being only a few hours by plane from Adelaide and Sydney, and having almost instantaneous contact with the rest of Australia and beyond by modern communication links.

I traversed Australia during part of one day, leaving Sydney in late morning and arriving in Perth that evening. There I was met by an old friend, David Ride, director of the Western Australia Museum, and by Duncan Merrilees, geologist at the museum. Duncan very soon became a new friend and a boon companion in the outback of northwestern Australia, which was to be my ultimate destination on this venture.

I wanted to go up into the northwest corner of Australia, 1,000 miles beyond Perth, for two reasons. One was to see something of the Triassic there, where John Cosgriff and other paleontologists had collected fossil amphibians that show close relationships with amphibians from the Lower Triassic of the South African Karroo. This was important to me because it was another link in the Gondwanaland story. The other reason was to have a look at, and perhaps make casts of, some large Cretaceous footprints exposed along the coast, which several years previously had been described as having been made by an iguanodont type of dinosaur. I had some suspicions about the tracks being those of an iguanodont, although they certainly were dinosaurian, and it seemed like a good opportunity to get some first-hand information about them. Whatever they were, they indicated that a big dinosaur had been wandering around in what is now northern Australia during late Mesozoic time; hence Australia would seem to have been connected with Africa and perhaps India, where large Cretaceous dinosaurs lived. Which means that it must still have been an integral part of Gondwanaland. Thanks to the kindness of David Ride and the museum, Duncan was to lead me there and show me what I wished to see.

We would go first to Derby, a little town at the head of King Sound on the eastern side of Dampier Land, and after that to Broome, another little town on the coast that bordered the west side of Dampier Land. We planned to make the trip to the north as soon as possible, particularly to be in Broome on May 26, the date of neap tide for that month. Broome enjoys extraordinarily high and low tides, the greatest tides in the world (I think) after those of the Bay of Fundy. The dinosaur footprints in which I was interested happened to be under water most of the time, being fully exposed only on the day of neap tide, and then for only about three-quarters of an hour. Therefore it was crucial that Duncan and I be there at the very time of day when the footprints were above water if we were to make casts of them. But for a while it looked as if we might not be able to get to Derby and Broome in time to carry out our work, because the office of the MacRobertson Miller Airline (a small domestic line) in Perth informed us that there were no places available on a plane for perhaps two weeks.

What were we to do? Duncan and I spent a frantic morning trying to work out possible plans to make the trip by car. But such a trip would have been a slow and arduous journey over dirt roads, with the chance that we might not achieve our objective in time to make casts of the footprints. By late afternoon I was feeling very discouraged at the prospect of ever seeing Derby and Broome, and that evening I went to bed with small expectations for the immediate future. The next morning, Bob Vincent came to our rescue.

Every good museum has a knowledgeable business officer, a man who knows a thing or two, and Bob Vincent was our man. He got in touch with the airline for us, and within a short time had made reservations on the plane leaving that evening. I enjoyed an outdoor lunch with David and Duncan in a beautiful park overlooking Perth. It was a nice setting; the shining modern buildings of the business center stretched to right and left below us, while beyond were the green hills of the western Australian coastline. There was nothing to worry about.

That evening about eight o'clock Duncan and I arrived at the airport with our luggage, ready for an all-night trip to the empty northwest corner of this great island continent. Our luggage was weighed and then we were asked to get on the scales, a procedure which I thought had gone out of style many years ago. We then walked out to our plane, an old-time DC-3, and when we entered in the back I immediately realized why we had been weighed, and why there had been some doubt about our obtaining seats on this flight to Derby and Broome. There were only four, or perhaps it was six, seats available at the very back of the plane; the rest of the fuselage was packed ceiling-high with goods. We were in effect riding an aerial freight carrier, whose business was to get the necessities of life to people in an out-of-the-way part of the world. And there were all kinds of necessities: crates of groceries, machine parts, goods of many varieties, and bags of mail. Duncan and I settled down for a long night—an up and down trip with stops to deliver items from the cargo in front of us.

It was just that. I dozed fitfully, every now and then opening an eye to cast a half-comprehending look at the mountainous cargo in front of me, and coming more or less to life when we came down for a landing. Early in the morning we were over Derby, and we descended through the long, slanting rays from a rising sun for our final landing. It was not, however, the final landing of the plane, because after it had discharged a large pile of goods on the Derby airstrip it took off once more, to visit, we were told, several cattle stations (ranches in American parlance) where still more deliveries would be made.

I will not go into detail about our work at Derby. It was the sort of paleontological reconnaissance that is useful and often fascinating to participants, but perhaps boring to others. Thanks to the kind help of Alan Ridge and Jim Coleman of Derby we were able to spend several days in the Erskine Range, about eighty miles to the east of town, where we poked around in the Triassic and found some fossil amphibians. We were in the outback, where there were few people but quite a number of wallabies, which would go bounding up the rocks if they caught sight of us. Also, as is so common in back-country Australia, there were billions of flies—flies that settled on our shoulders and backs so that our shirts were literally black with them. We wore broad-brimmed hats with nets hanging down to keep the flies off of our faces and out of our eyes, ears, and noses. Many Australians arrange a fringe of corks suspended from threads from their hat brims. The dangling and swinging corks seem to help keep the flies away. The flies were a nuisance but I learned to live with them. I also learned to live with the spinifex grass, which grows in clumps of sharp, porcupinelike, spiny leaves. These were merely some of the local discomforts—every environment has its own discomforts, to balance the comforts. We ignored them and went ahead with our work, which to me was new and, needless to say, of great interest, and I gained my first-hand acquaintance with the Lower Triassic rocks of western Australia.

Also I became acquainted with the grotesque baobab trees which we encountered during our wanderings east of Derby. These strange trees can best be described as having comparatively short but enormously inflated trunks, with heavy, crooked branches projecting out to form broad, scraggly, spreading crowns. One such tree by the road near Derby was so very large that in former years the partially hollow trunk had been used as a sort of lock-up for petty transgressors of the law. This old veteran was one of the last things I saw as we made our final trip from the field into Derby, where we would pack the fossils we had collected preparatory to our departure for Broome.

The early morning flight to Broome was a short hop and we landed in time for breakfast at the Broome Continental Hotel. We were taken there by John Tapper, harbormaster of the Harbor and Lights Office, and after breakfast he picked us up for a short ride to the beach where we would see the dinosaur tracks.

The hotel at Broome was a rather unprepossessing structure as seen from the outside—a long, low, one-story building with galvanized-iron siding, a tin roof, and a verandah along the front. First appearances can be deceiving; closer acquaintance with the hotel showed it to be in the form of a hollow square, composed of four long, narrow wings surrounding an inner courtyard or garden, in which there was a pleasant terrace shaded by trees and a green lawn edged by beds of flowers. Our room, in one of the wings, opened onto a long inner verandah, as did all of the rooms, shielded from the inner garden and the rooms on the opposite side of the court by a latticed screen. The verandah afforded welcome, cool shade to supplement the amenities of the garden beyond.

The wing opposite our sleeping quarters was occupied by the dining room and kitchen and other public rooms, while at right angles, connecting the facing sleeping rooms and the dining room, was the wing occupied by the bar. It was a big bar and it was enthusiastically patronized by Aussies from the town and from the outback.

The men who live in the Australian outback are necessarily very rugged characters; indeed it might be fair to say that they are similar to the old-time American cowboy and roustabout, only more so. I think I would go even farther and say that the outback Australian makes the American westerner (who prides himself on being something of a tough guy) look like the participant at a Sunday-school picnic. Such being the case, I found the bar to be noisily vibrating with the activity going on inside it. It vibrated during the late afternoon and into the evening, and on that first night I wondered how much sleep Duncan and I would get. I needn't have worried; promptly at ten o'clock the bar shut down, and the night was quiet and peaceful as only a night in such an isolated place can be. The Australians certainly have very strict closing times for their alcoholic emporiums.

But to get back to the dinosaur tracks. We went out to Gantheaume Point, where the lighthouse is located, and to nearby Riddell Beach, at both of which places we saw footprints. They were large, three-toed tracks, obviously made by a bipedal dinosaur, and they could be discerned be-

Large, three-toed tracks, obviously made by a bipedal dinosaur, could be discerned beneath the swirling tidal waters.

neath the swirling tidal waters that were washing over them. The water covering them was only a few inches in depth; tomorrow they would be completely exposed, and then we would make our casts.

Now we had to discuss ways and means. We had brought from Perth a sufficient quantity of casting compound for the task. This was a material used by dentists—I can't remember its name—that mixes quickly, sets within a very few minutes, and makes a flexible cast. I think we must have purchased enough of this casting compound to keep all of the dentists in Western Australia supplied for several months on end. The idea was to mix this material and pour it into several dinosaur footprints when they were exposed and dry. It would be necessary, however, to extend the cast (one might more properly call it a mold) above the footprint, thus making a base to give it strength, and to afford a means by which it could be extracted when it had set. How were we to contain this part of the casting, which would project above the level of the rock into which the footprint was impressed? We would not have much time—only half an hour or so. And the wall which would contain the upper part of the cast would necessarily have to be flexible, to fit the contours of the rock surface. At this point Mr. Tapper came up with a very good suggestion; namely that we use a coil of heavy rope, placed around each footprint. The rope could be cut to the proper length and be ready for immediate use. Then it would be only a matter of a minute or so to put it in place. We all agreed to this, and made our plans accordingly.

Our discussion had taken place on the rocky floor of the beach, a surface formed by the hard Broome sandstone of Cretaceous age, constantly washed clean by wave action. There was no sand to hamper us. Behind us rose craggy cliffs of red, cross-bedded sandstone, perhaps fifty feet or so in height. And during our conference below these cliffs, a pair of black, fork-tailed frigate birds rode the thermals that came in from the ocean and swept up over the cliffs. It was marvelous to watch them, hanging almost motionless in the air as if suspended from strings. Then, when we returned from the shore into town, we were greeted by the sight of a great flock of white cockatoos settling down onto some vacant ground beyond the widely spaced houses.

At dinner that evening we sat at a table with Edgar Truslove, an official of the MacRobertson Miller Airline, and he became more than a little bit interested in our project. We immediately enlisted him as one of our helpers in the casting project. Also, Edgar Tapper, brother of John Tapper, planned to work with us.

The next morning we all went out to the beach, where we selected the prints that were to be cast (we could see them beneath the water) and practiced our procedures. It was a sort of preliminary drill for the real thing. Then in for lunch, after which we returned to the shore. The tide would be at its lowest stage in midafternoon, so we got ready for action.

Finally the water went down at Riddell Beach, and we began work on the best footprint of a three-track trail there. The coil of rope was placed around the track and held there by some rocks placed around it. The casting compound was mixed in a plastic pail and poured onto and above the track. Within about five minutes it was sufficiently set so that we could carefully extract it from the footprint.

After this we galloped over to nearby Gantheaume Point, where we repeated the process on three footprints there. All of the castings turned out successfully, so we felt very good about the outcome, completing our work just as darkness overtook us and the tide started to wash in over the dinosaur tracks. (I had in between times made photographs and sketches of the trackways, so we would have information as to how the footprints we had cast related to the complete sequence of tracks.)

We went back to Mr. Tapper's house, where we carefully placed the casts in some boxes that Mr. Tapper with our aid had constructed to hold them. Then we were ready to return to Perth. Fortunately we were going back on a regular passenger plane, and Mr. Truslove was going with us. He saw that the box with the casts was properly stowed on the plane before we took off.

Back at the Western Australian Museum in Perth the tracks were unpacked, and the news and television people came in to interview us about them. They liked the idea of a story concerning dinosaurs that lived in Western Australia more than 60 million years ago. Three years later Duncan Merrilees and I published a little paper in the *Journal of the Royal Society of Western Australia* describing the prints. They had been made, we decided, by a big carnivorous dinosaur, as indicated not only by their size and shape, but also by the imprints of claws at the end of each toe. We named the tracks *Megalosauropus broomensis*—"the feet of a megalosaur from Broome."

The megalosaurs were gigantic carnivorous dinosaurs which, according to previous records, lived in Europe and in the African segment of Gondwanaland during late Jurassic and early Cretaceous times. Now we had found indications of early Cretaceous megalosaurs in Australia, for the

Duncan Merrilees pouring casting compound into a dinosaur track, with Edgar Truslove assisting. With them are John Tapper, Harbormaster at Broome, and his brother Edgar.

Broome sandstone had been independently dated as being of this age. One more bit of evidence had been added to the history of Gondwanaland. In this last period of the Mesozoic it would appear that the ancient supercontinent was still intact, at least to such an extent that there were avenues of migration whereby giant carnivorous dinosaurs could wander between Australia and Africa (perhaps by way of Antarctica) and beyond. I might add that when I had been in Brisbane I had seen the partial skeleton of another early Cretaceous dinosaur that had been found in Queensland. So the picture of Gondwanaland as a vast range for the distributions of ancient reptiles was becoming ever more fully delineated, not only in space but in time.

There is little else for me to say about my Australian adventures. I flew back from Perth to Sydney, where I spent a week or so, devoted largely to some one-day field trips with Harold Fletcher of the museum, to see more of the eastern Australian Triassic. I crossed to New Zealand and spent some time there, looking at fossils in museums and rocks in the field. Then I returned home.

15. ANTARCTICA

One day, early in 1968, the telephone in my office rang and when I answered it a voice at the other end informed me that he (the owner of the voice) had just returned from Antarctica, and he had what he thought was a fossil bone. He was at the Institute of Polar Studies at the Ohio State University in Columbus. If he brought the specimen to New York would I be willing to look at it and give an opinion? Would I? I certainly would, particularly since he said it came from Triassic beds about 400 miles from the South Pole. This was more than exciting news—it was sensational!

For years I had been hoping that some fossil vertebrates would be found in Antarctica, especially Mesozoic vertebrates, and here was something which, if valid, would be our first evidence of a land-living animal of significant geological age from the south polar continent. Its bearing upon Gondwanaland and upon the whole theory of plate tectonics and drift might be of the utmost significance. I awaited the arrival of the bearer of the bone with a mixture of impatience, curiosity, and some foreboding. Would the fossil really be a fossil bone?

In a day or two he appeared—a young man named Ralph Baillie. He opened the parcel that he had brought with him, and from within an ample layer of cotton he extracted an object about four inches or so in length. *It was a fossil bone!* Furthermore, it obviously was a fragment of a lower jaw of a labyrinthodont amphibian, the labyrinthodonts being those bony and generally large amphibians that lived during late Paleozoic and Triassic times and that, for some millions of years before the reptiles arose from labyrinthodont ancestors, had been the uncontested rulers of the land. I knew it was a labyrinthodont because of the heavy sculpturing on the outer surface of the jaw, and I knew it was a jaw because near one end of the bone was a smooth, semicircular depression—the articulation for the quadrate bone of the skull, permitting the lower jaw to rock up and down. The fossil fulfilled my fondest hope!

Ralph told me that he had been in the field with Peter Barrett, a New Zealand geologist who at the time was associated with the Polar Institute. On December 28, 1967, they had been exploring the Lower Triassic Fremouw Formation on the slopes of Graphite Peak in the Central Transantarctic Mountains when they found the specimen. Even to have seen the fossil was a feat of remarkable eyesight, combined for sure with quite a bit of good luck. The jaw was in place in the rock—a coarse sandstone—with the smooth and not very diagnostic inner surface exposed. At best it was a difficult thing to see, and considering that Peter Barrett and his assistants were not specifically looking for fossils they deserve the greatest credit for having found the bone. If they had not been so sharp-eyed and so lucky on that December day there probably would be no story to tell here.

Ralph was willing to leave the specimen with me for identification and study. And one of the first things I did was to take the fossil down to Princeton University, to show it to my friend Don Baird. Don is one of those fortunate paleontologists who can see the true relationships of even the most fragmentary and frustrating fossils. I was sure I knew what the fossil was, but I wanted corroboration from Don; if he agreed with me, I knew there could be no doubt as to the nature of the bone. Don looked at the fossil and he agreed. Which made me feel very good indeed.

Here we see how chance often looms large in scientific discovery and research. It was only by a lucky chance that Peter Barrett and Ralph Baillie happened to walk over the very spot on Graphite Peak where the fossil was exposed, a spot where it had rested for many millions of years. It just so happened that Ralph Baillie called me about the fossil—he might have gotten in touch with any one of a dozen other people. But he didn't; he called me, and thus began my greatest paleontological adventure.

An amusing sidelight. When Barrett and Baillie arrived at Christchurch, New Zealand, the staging area for all American flights to and from Antarctica, the quarantine people immediately seized the fossil bone and impounded it in a refrigerator. The explorers returning from Antarctica had made the mistake of writing "fossil bone" on their declaration, instead of "rock." Now, the fossil was mineralogically a rock, completely inert, but that word "bone" got the government officials excited. There is a strict rule about the importation of organisms into New Zealand, and with good reason, so the quarantine people were taking no chances with a bone, no matter its age. Consequently the fossil was kept locked up in the refrigerator, at a temperature probably much higher than it had been exposed to in Antarctica for many years, until it was time for Barrett and Baillie to come on to the States. The fossil was then ceremoniously removed from the deep-freeze and returned to its discoverers.

I worked on the fossil and prepared a description. In the August 2, 1968, issue of *Science* the paper appeared, a joint contribution by Barrett, Baillie, and myself entitled "Triassic Amphibian from Antarctica." The announcement caused a considerable amount of interest in the scientific world, and came to the special attention of various people in the Office of Polar Programs of the National Science Foundation.

In December of that year I attended the annual meeting of the American Association for the Advancement of Science in Dallas, where I presented a report on the fossil. While there I was talking with some of the NSF people, and I said in effect: "Here is a fossil that was found by accident, by geologists working on another field problem. Where there is one there should be more. Why not send a party to Antarctica specifically for the purpose of finding Triassic amphibians and reptiles. I think something should come of it."

"Fine," they said, "let's send down a party under your leadership."

"But I'm too old," I said, "and furthermore I am getting ready to retire in a year or so."

"Don't worry," was the reply. "You go, and we'll take care of you."

I girded up my loins, so to speak, for an exploring trip to Antarctica—to the cliffs and rock exposures of the Transantarctic Mountains less than 400 miles from the South Pole. Before I left for Antarctica there were a few things to do. For one thing, I had to pass the rather extended physical examinations required of potential Antarctic scientific workers. I consequently spent a day at the Naval Hospital on Long Island, undergoing all sorts of tests, which I passed to the satisfaction of the various doctors involved.

For another, I had to select the people who were to work with me in Antarctica. I hoped to have Dr. James Jensen of Brigham Young University in Provo, Utah, as a member of the team. I got in touch with Jim and he accepted with alacrity, which made me happy, for if we were to have success in the field it would depend as much as anything on Jim's devoted application and know-how. Jim had spent many years at Harvard as a right-hand man to Al Romer, after which Jim had developed a very active program at Provo. He was a whiz in the field. I also hoped to have Gil Stucker of the American Museum with us and Gil was especially anxious to go, but the doctors at the Navy hospital would not pass him. That stunned both of us. Gil was and is a large and vigorous man with much outdoor experience, as well as years spent in the field collecting fossil vertebrates. The medical men were adamant, and their refusal to accept Gil was a real setback to our plans.

Another member of the team was to be William J. Breed, geologist at the Museum of Northern Arizona, again a man with much geological and

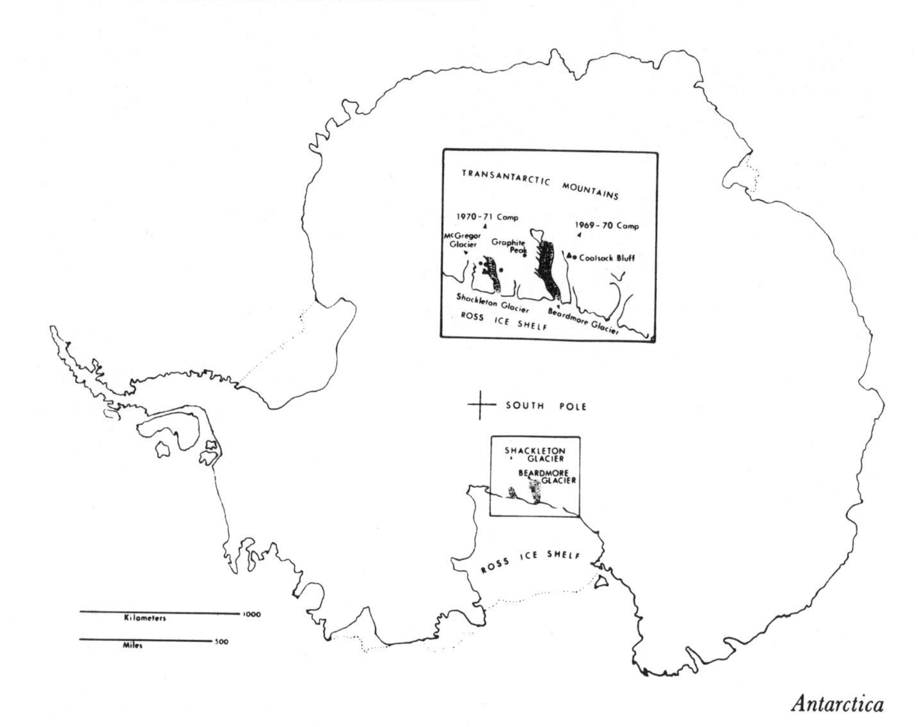

Antarctica

outdoor experience. Since we could not have Gil, we recruited a geology student from the University of Arizona, Jon Powell.

We were four, but by no means the full complement of the team that was to work in Antarctica during the coming field season. Our efforts were to be part of the general program of the Institute of Polar Studies at Ohio State (as well as the National Science Foundation), and the institute had plans for a field effort that involved much more than fossil vertebrates. One factor determining this had to do with logistics. It is very expensive to put a field party down on the ice in the middle of Antarctica; therefore the party might as well be of some size because the supply planes (Hercules C-130s) can transport and take care of big parties more efficiently than small ones. Beyond this, the Institute of Polar Studies was and is dedicated to comprehensive research on polar problems, so that any opportunity to study Antarctic geology on a wide front is an opportunity to be seized. Therefore plans were made for a rather large field party to work in the region where we wished to look for fossils, studying various aspects of the geology in that particular area.

The leader of the whole group was to be David H. Elliot of Ohio State, a man with much Antarctic experience. I became acquainted with David during the months before we went to the Ice, and this acquaintanceship quickly ripened into a close friendship that has continued through the years. Indeed, David and his family have become dear friends of our family.

David is a displaced Englishman, who has now spent a good many years in Columbus, to become at the present time the director of the Institute of Polar Studies. He is a fine scholar and an experienced geological field man. Moreover, he is a leader with good judgment and understanding. Needless to say, I soon realized that we would have excellent direction at our field camp, which gave me much preliminary confidence in what we were going to attempt to do. David is especially interested in volcanic rocks, and there would be volcanics in abundance where we were to work.

In addition to David and ourselves, the field party was to include William Gealy and John Splettstoesser, administrator of the Ohio State Polar Program; Jim Collinson working on stratigraphy; Henry Brecker, gravimetric studies; Paul Tasch and his assistant, Dietmar Schuchmacher, involved with conchostracans, which are small, fresh-water fossil invertebrates; James Schopf, Leon Lambrecht (of Belgium), and Josef Sekyra (of Czechoslovakia), paleobotany; Isak Rust (of South Africa), general geological studies; Mike Peterson, paleomagnetism; John Gunner, Precambrian geology; Don Coates, glacial geology; John Mercer, studies of geologic age; and Roy Cameron and his assistants, Roger Hanson and George Lacy, microbiology. It was to be a considerable and varied party with a wide range of scientific experience and viewpoints—all of which promised to make things interesting.

We would have Navy support: three helicopters with five pilots and nine or ten mechanics, hopefully to keep the flying machines in good running order, a cook and his assistant, a radio man, a medical technician, and a handyman or two.

Margaret and I had to prepare for a move from our New Jersey home of thirty years to our new home in Arizona. In 1968 we had begun the construction of a house at Flagstaff, on land that we had purchased the previous year. The house had been completed for us, largely while we were *in absentia,* and now we were ready to make the move. I had not technically retired from the American Museum; that was to come during the following February. The administration wished me to continue on the active staff until after the Antarctic venture, in order that the museum could be an institutional participant in the work. I did, however, have vacation time coming, and some accumulated leave, so we planned to depart from the east in July and drive to our new home, where I would remain until it was time to go south to the Ice. That would be in October.

It was with mixed feelings that we vacated the house in which we had spent so many years, the house in which all of our children had grown up. Naturally, there were strong ties to the old place, but such feelings of sadness that we had in leaving it were alleviated in part by the excitement and promise of looking forward to a new life in our new home.

During the summer we settled ourselves in the new house, doing the thousand and one things that have to be done when a move is made. As the late summer blended into the lovely autumnal days of the Arizona high country I began to prepare myself, physically and mentally, for the long trip to the other side of the earth. It was hard to leave. It was hard to have to say farewell, even temporarily, to the house in which we had invested our plans and hopes for future years, to the tall, straight ponderosa pines that surrounded us, to the golden aspen groves on the flanks of the San Francisco Peaks that rose behind our house, and to the warm sunshine of dreamy autumn days.

Shortly thereafter it was hard to say farewell to Christchurch, New Zealand, now enjoying the height of its spring weather. Every day for almost a week at Christchurch we would rise before dawn, have an early, specially prepared breakfast at the hotel, and then drive to the airport. There we would don our Antarctic gear (which was required on the theory that our plane might come down in the middle of the stormy sea between New Zealand and Antarctica—a less than cheerful thought) and then we would sit for an hour or so, only to be informed that the flight had been cancelled for that day. The weather in Antarctica was very stormy, so we could not take off from Christchurch if there was a prospect of landing difficulties in Antarctica.

After each aborted flight we had the day to ourselves, so we made little trips in and around Christchurch to see the sights. Well do I remember a day spent in the botanical gardens, which are very fine, wandering across green lawns and viewing everywhere the beauty and profusion of the spring flowers. I felt very sorry for myself, thinking of the necessity of leaving this arcadian scene for the bleak glaciers to which we were going.

On our fifth morning trip to the airport we once again put on our bulky Antarctic clothing and made ready for the trip, only to be told, as before, that the flight was scrubbed. This time, however, our respite was not to last until the next day; we were told to come back immediately after lunch, for we might make the trip after all. Back we went and about two in the afternoon we boarded the plane—a gigantic Hercules C-130 for our flight south. The Hercules has revolutionized scientific work in Antarctica. It is a big plane that can carry many people and tons of cargo. It has four propeller engines, and can fly with one or two or (I have heard) even three engines nonfunctional. For Antarctic work the planes are equipped with huge skis so that they can land on the ice. With this aircraft, the problems of supply have been solved. The Hercules can set a field party down on the ice with all the necessary supplies and equipment for two weeks, a month, or two months. The scientists working in Antarctica can devote their at-

The planes are equipped with huge skis for landing on the ice.

tention to the tasks ahead instead of to a struggle to stay alive, as it was in the old days.

The interior of our Hercules looked like an oversize moving van, with piles of supplies and equipment rising from the floor to the two-story ceiling. There were seats along the sides of the fuselage, where the human occupants made themselves comfortable as best they could. We strapped ourselves down and got ready for the take-off. (It should be explained that the usual wheels are still retained on the ski-equipped plane; they are let down below the skis for take-offs and landings on regular airstrips.)

Off we went, and for about four hours droned along on our southerly course. Suddenly the plane began to bank into a 180-degree turn, and we were headed north again. We had reached the halfway point, the point of no return, and word had come through on the radio that landing conditions had worsened. Back to Christchurch we flew, landing about nine-thirty in the evening. This time, however, our sojourn in Christchurch was to be a short one. By four o'clock in the morning I was up again and soon thereafter was on the way to the airport in company with a number of bleary-eyed Antarctic hopefuls.

We were off the ground by six A.M. and had a steady eight-hour flight to Antarctica. As we came in over the Hallett Coast I had my first glimpse of the south polar continent, and an ominous glimpse it was. Below us was a huge, ice-covered terrain with steep cliffs descending to the shore, the entire scene alternately hidden and revealed by swirling clouds. We came down at Williams Field, an airport on the ice to serve McMurdo Base, the American staging area for Antarctic work. As we stepped out of the plane we were hit by a cold, below-zero wind that slashed at our exposed faces, causing us to hunch down into our parkas as much as possible, red parkas for the scientific workers, green parkas for the Navy men.

There I had my introduction to a dichotomy among American Antarctic personnel. The scientific people are under the jurisdiction of the United States Antarctic Research Program, and are generally known as Usarps. The Navy provides logistical support; all in all it requires about a dozen Navy men for each Usarp. There is, however, a second dichotomy. The Usarps are ranked as a species of protem Navy officers and thus are set apart from the larger body of enlisted Navy men.

As soon as we had collected our gear we got into trucks and were hauled off to our quarters at McMurdo Base. Little did I realize as we bumped across the ice that my quarters would be my home for three full weeks, but that's the way it was. Antarctica was enduring some of the worst spring weather since the days of Scott's fatal expedition of 1911–1912, which meant that we were frequently confined to quarters, and when not so confined, limited to short field trips. In Antarctica one bends with the weather.

Being confined to quarters at McMurdo Base was no great hardship, for it was a comfortable place in which to wait out a spell of stormy Antarctic weather. There were solid buildings, including a large mess hall in which we were served ample and very good food. There was a bar with a variety of liquid refreshments, and adjacent to it a room where every evening we would be entertained with movies. Some of the most wonderful movies we were privileged to see were "westerns," filmed in Czechoslovakia (or perhaps it was Yugoslavia) and based upon the stories of Karl Mai, a German writer who enjoyed immense success in Germany early in this century, despite the fact that he had never been to North America. The films had a most marvelous mixture of Plains and Pueblo cultures, with Algonquin canoes thrown in for good measure, along with cardboard saguaro cacti, pueblo dwellings complete with balconies and some teepees in the front yard, and old western towns (one ostensibly Santa Fe) with solid stone houses, and not far away the shore of the sea. How these films got to McMurdo I shall never know. I do know that Bill Breed and I itched to get our hands on one and bring it back to Arizona, where we could show it to our Hopi and Navajo friends. Alas! That was never to take place.

My own quarters were very cozy; a room in a hut, the room being just the length of a bed with about three feet of space in addition to the width of the bed. The bed was a double-decker bunk, and I had the upper level with about two feet of head room. There was no window, and when the light was off it was as dark as the bottom of a coal mine. This was good because we were in Antarctica during the austral summer, when there are twenty-four hours of daylight. So I could go to bed, turn off the light, and think it was really nighttime.

With the lights on I spent many comfortable hours relaxing and reading. Books were available at McMurdo, and among others I found one about the Greeley expedition to the Arctic, written interestingly enough by a former next-door neighbor in New Jersey, Alden Todd. This book recounted in terrifying detail the sad tale of the Greeley expedition, involving failure, death, and even cannibalism. To compound the terror, a doctor from New Zealand, a member of the ice-training team of instructors, gave us a lecture one evening on frostbite, with slides that were all too realistic. Furthermore, it so happened that not long before we departed for the Antarctic I had seen a film on TV—*Scott of the Antarctic*—showing the trials and sufferings and the deaths of some of the men on that expedition. Outside the wind howled and the snow was driven through McMurdo Base in horizontal gusts. That was the stage when I was ready to turn around and go home. Actually, when we finally did get into our field camp, I enjoyed Antarctica very much. It was not nearly so terrifying as I had expected it to be.

When the weather permitted there were active employments that occupied some of our McMurdo time. To begin with, there was a required ice-training course—a strenuous exercise that lasted for several days. Under the direction of some New Zealand mountain climbers (including the doctor who had frightened us with his pictures of frostbite victims) we slid down ice slopes, leaped into crevasses at the ends of ropes and were pulled out, cut steps up icy cliffs, and in general went through experiences which I, for one, hoped would never be encountered in the field. But we had to be ready for emergencies. I shall never forget my clumsy efforts, kindly supervised by Allan Cookson, a dentist from Christchurch (with whom I established a lasting friendship) during that memorable ice course. Manfully (I hope) did I whack away at an icy cliff with my pick, cutting steps up which to climb, while the sweat poured off my forehead and ran down inside my goggles, effectually to blind me, so that everything I tried to do was seen through a shimmering film of salty water. We were told not to perspire, but on a calm Antarctic day I would sweat like a plowhorse; I couldn't help it.

Then there was field equipment to be checked and various other chores. Finally, we managed to get in some one-day trips here and there, and this helped to fill the time, as well as to give us some much-needed experience.

The one-day field trips out of McMurdo were adventurous enough. The first, which also was my first day in the field, was by helicopter to some Permo-Triassic exposures on Allan Nunatak, perhaps seventy miles from McMurdo. (A nunatak is the tip of a mountain projecting through a glacier usually thousands of feet in thickness.) As we reached our destination I could look down and see the snow blowing in blizzardlike clouds across the ice. We should never have landed, but we did. Then we had a miserable day, trying to hike across the landscape and see some geology, and contend with the wind. I was knocked down by the wind three times on that day, and in addition my cheeks began to get frostbitten, but Jim Jensen noted my predicament and warmed my face with his hands.

Late in the afternoon we got down off the nunatak and waited for the helicopter to come and pick us up. It did not come, and it did not come, and finally we decided that we would have to camp out for the "night." (In Antarctica one is always left with emergency gear.) Just as we were making preparations we heard the helicopter in the distance and soon we saw it. We jumped up and down and waved our arms, but the helicopter circled, evidently without seeing us in our bright red parkas, and flew away. There was a general sinking of hearts, but our disappointment did not last long for in a few minutes the helicopter was back. This time we were seen, the helicopter came down, we scrambled on board, and soon were on our way back to the warmth of McMurdo.

Four days later we got ready for another trip to Allan Nunatak but the weather was against us. There were two succeeding aborted attempts to reach it, and then, seven days after our first trip, we did get there, Bill Breed, Jim Jensen, Jon Powell, and myself—we being the vertebrate paleontological contingent, and in addition David Elliot, Jim Schopf, Jim Collinson, and Isak Rust—the geological-paleobotanical group. We traveled in two helicopters.

It was an interesting day. Isak and I found some large fossil logs contained within a "sparkling" sandstone, all of which reminded Isak of the Molteno beds of the Karroo sequence. Then, high on the nunatak, we found some very fine examples of *Dicroidium,* a fernlike fossil plant quite characteristic of the Upper Triassic sediments of Gondwanaland. Late in the afternoon we boarded our helicopters feeling good about what we had seen. Certainly the fossil plants were most convincing; they told us of an ancient Gondwanaland with a tropical or subtropical climate, and they told us of Antarctic-African connections.

As we were flying back to McMurdo, happily contemplating our day in retrospect, we were suddenly horrified to hear over the radio that there had been a helicopter crash at the foot of the Wright Dry Valley. The helicopter, as we learned later, had lost power and had been forced to come down with a spinning but powerless rotor, so there was little control over its descent. It landed on a steep mountainside, slid down the slope for several hundred feet, and burst into flames. Two people were killed, the others were injured. The pilot, although painfully burned, had walked fifteen miles up the Wright Valley to a little camp where a couple of glaciologists were studying the Meserve Glacier. He knew they had a radio, as is required at all Antarctic camps, and from there he sent out a Mayday or distress signal. And that is what we heard.

(It should be explained that the Wright Valley is one of the very strange Antarctic valleys devoid of snow and ice—for what reason no one is certain. Glaciers pour down on the sides of the valley from the heights above, but the valley floor is a bare, rocky surface.)

Our helicopter immediately changed course and flew toward the camp at Meserve Glacier. As we flew in we could see the pilot of the downed helicopter waving to us. We landed, and Jim Jensen, Isak, Jon, and I piled out as fast as we could, the injured pilot climbed in, and off went the helicopter to the rescue. It was a tragic ending to our day.

The two men at the Meserve camp made us welcome and we unrolled our sleeping bags and prepared to spend the night. Just as I was comfortably drowsy there came the chop-chop-chop of a rotor, and we knew a helicopter was coming to pick us up. There was a mad scramble while we donned clothes and rolled up our sleeping bags, and we just managed to be ready when the machine landed. In this fashion we concluded a day that was geologically satisfactory to us but very sad for all at the McMurdo Base. We learned when we arrived back that the people killed were a geologist and a New Zealand photographer. The following day the whole base was in a state of shock.

On November 22, three days after our second trip to Allan Nunatak and the helicopter crash, we finally did leave for our field camp. We had planned to go on the previous day, but once again, as had so frequently happened, the weather defeated us. We took off in a Hercules at about ten-thirty in the morning and had a magnificent flight along the Transantarctic Mountains to our destination. The day was bright with sunshine, the mountains loomed majestically before us, and the great glaciers below glistened in menacing detail. It was difficult, looking down on the immense

Our camp consisted of four Jamesway huts, these being portable versions of Quonset huts, with arched wooden frames, over which were stretched insulated fabric coverings.

Beardmore Glacier, to comprehend how Scott and his party had managed to journey along the length of that ice river with its thousands of deep crevasses.

For me the beauty of the flight was marred to some extent by the comments of David Elliot as we flew across the mountains. David and I, as co-leaders of the expedition, were riding in the pilot's cabin and marking the progress of our flight on the maps we had. David would lightheartedly point through the cabin window to a vast, absolutely frightening cliff ahead, and tell how we could land a helicopter on the pointed peak above it, and then make our way down to the exposures we might want to see. I viewed the prospect with something less than enthusiasm. Luckily for me, we never had to do this, for we found fossils on nice, gentle cliffs, readily accessible to human beings rather than mountain goats.

We landed at our camp about one-thirty in the afternoon and spent the rest of the day getting established. The camp had been set up for us by a contingent of Navy seabees, who had flown in several days ahead of our party. It consisted of four Jamesway huts, these being portable versions of Quonset huts, with arched wooden frames, over which were stretched insulated fabric coverings. The huts had wooden floors, and each hut had an oil heater, so it was comfortable enough. It must be admitted, however, that the comfort zone was at about sitting level; water spilled on the floor of the hut would quickly freeze, while at the very apex of the hut the temperature was perhaps well up in the eighties. Nevertheless, we had no reason to complain.

One of our four huts was for the Usarps, one for the Navy personnel, one was a mess hut, and one was a radio and first-aid shack. In addition there was a little plywood structure housing a generator, so we had some electricity at hand. Not that we needed it for light, because, as has been noted, we were enjoying twenty-four hours of daylight.

We were camped on a large ice field known as Walcott Névé, principally because this was a good place for our supply planes to land. It was our intention when we had helicopter support to fly across the Beardmore Glacier to Graphite Peak about sixty miles away, to begin looking for fossils. That was where Peter Barrett and Ralph Baillie had found the one small fragment of amphibian jaw on which this whole massive effort was based.

We had arrived three days ahead of the helicopters that were to give us wide-ranging support for our geological and paleontological explorations. On the first morning in camp we looked around for things to do. I thought I had some things to do in camp, getting myself more firmly established and writing up some notes. David Elliot, Jim Schopf, and some of the others thought that it would be fun to go over to a nunatak known as Coalsack Bluff, about five miles from camp. They went mainly because it was there, a small mountain at one end of a little range terminating at its opposite end, about twenty miles away, in a somewhat higher eminence named Mount Sirius. I think another motivation for their excursion was an argument as to whether the sedimentary rocks exposed at Coalsack Bluff were of Permian or Triassic age. They were both, the Permian Buckley Formation at the base of the bluff with the Lower Triassic Fremouw Formation above. (A geological formation is a distinctive rock unit that can be mapped.) The top of the bluff consisted of dolerites, which are volcanic rocks. We had some heavy-duty motor sleds in camp, so David and his curious-minded colleagues took one of these, hitched to it a Nansen sled (a light, many-purpose sled specially designed for polar work), and departed for Coalsack Bluff.

About lunchtime they returned in a high state of excitement. They had found Triassic fossil bones in the Fremouw cliffs of Coalsack Bluff, and they brought back two tiny samples to show me. They were fossil bones all right—but of what? We ate a hurried lunch, and then a big group of us went back to Coalsack Bluff for the afternoon. As we explored the low sandstone cliffs we repeatedly found fossil bones exposed in the rocks. Nothing was articulated but there were numerous isolated bones waiting to be excavated. Obviously this was a place to be carefully worked. We went back to camp in a state of euphoria, our heads full of plans. That evening I wrote in my notebook: "The implications of these bones are of significant import—especially as they bear upon Drift. There can be no doubt but that Antarctica was once in contact with other continental blocks."

The next morning we began working at Coalsack Bluff, going there by sled and taking some food along so that we could spend the day. Bones were to be seen one by one along a length of several hundred yards at least, and it was our task to get them out of the rather hard sandstone in which they were contained. It was a channel sand, deposited by an ancient Trias-

Bones were to be seen along a length of several hundred yards at least, and it was our task to get them out of the rather hard sandstone. Jim Jensen found Lystrosaurus *during the course of the afternoon.*

sic river, and it contained the remains of animals that had fallen into the stream to be covered by the aggrading sand or of dismembered carcasses that had been washed downstream, the bones breaking away to be separately buried. For us to recover the fossils there would be much wielding of hammer and chisel.

The work on our excavations was complicated by reason of our heavy clothing, particularly by the necessity of wearing mittens. It was dangerous to have one's hands uncovered for more than a minute or so at a time; consequently, we had to try to hold our chisels in heavily gloved fingers and to pound away with a clumsily held hammer. Control was difficult under such circumstances.

David Elliot (left) and the author looking at a Lystrosaurus *bone embedded in a channel sand which had been deposited by an ancient Triassic river. It contained the remains of animals that had been washed downstream, the bones breaking away to be buried separately.*

There was also the problem of bandaging the larger specimens. What does one do in a climate where water turns into ice so rapidly as to interfere with the setting of plaster? For that matter, how does one manage a supply of water to begin with? Jim Jensen solved the problem (and this was but one of many problems that he solved) by the novel technique of heating a mixture of beeswax and paraffin over a little alcohol burner, and then quickly pouring the hot mixture on narrow strips of burlap which had been wrapped around the fossil. It worked. Plaster was used sparingly during the following field season, but it required some special means for heating water and applying the plaster.

The day was spent in excavating and preserving fossils as was the next day. On the third day following our discovery of fossil bones the helicopters arrived—three of them. They flew in from McMurdo, accompanied by a Hercules which acted as a sort of shepherd for the whirlybirds. The helicopters had been fitted with auxiliary fuel tanks for the 400-mile trip, which was three or four times their normal range. They made the trip without incident, and it was quite a thrill to see them come in, one after the other, to settle down on the ice. Once they had safely landed, the Hercules left for other duties.

The next morning we arose with visions of extended reconnaissance trips by helicopter to look for fossils, and we decided to make a trial run, so to speak, by a survey of the mountains near Coalsack Bluff. After breakfast Bill Breed, Jim Jensen, and I boarded one of the helicopters for a short flight to our fossil locality, the rotors were set in motion, and we were ready to take off. Just then a radio message came in, saying that Larry Gould and Grover Murray were arriving in a Hercules for a brief visit, and would I please disembark to meet them?

Dr. Laurence M. Gould was and is the Dean of Antarctic scientists. He was second in command on the Byrd Expedition of 1929, and had come back to the Ice in part as a celebration of forty years of Antarctic exploration and research. Dr. Grover Murray was a key figure in the National Science Foundation polar research program. Of course I was anxious to meet them.

I got out of the helicopter and stood aside to watch it take off. John Splettstoesser was standing with me. The helicopter began to rise when suddenly there was a loud bang, pieces flew off in various directions, and the helicopter dropped down onto the ice with a thud. I started to run toward the machine, but John grabbed me and yelled, "Stay back! It may explode!" At that same instant the door of the passenger compartment slid back and Bill and Jim tumbled out, while up front the pilot and copilot likewise were taking their leave. The men all ran clear of the helicopter, and then we stood there to see what would happen. Nothing happened. The helicopter did not explode, nor did it burn, although there was a lot of smoke. It was a sorry sight with its tail askew, its skids crumpled, and its sides blackened by smoke. Luckily nobody was hurt. The only damage beyond that suffered by the helicopter was a bent rim on the lens of Jim's camera.

About five minutes after this excitement the Hercules landed, and the first sight that greeted the eyes of Larry Gould and Grover Murray was a very much disabled helicopter surrounded by a mixed crowd of Usarps and Navy folk.

That was the end of our helicopter support for about three weeks. It seems that the drive shaft to the tail rotor had snapped (perhaps it had suffered crystallization of the steel owing to the cold temperatures of our camp). The shafts of the other two machines were taken out and shipped to New Zealand for inspection and analysis. The damaged helicopter was out of service for good; eventually it was dismantled, loaded on a Hercules, hauled to McMurdo, and shipped out of Antarctica. That left us with two helicopters for the rest of the season, which drastically cut down the use of the machines. One helicopter always had to be kept in reserve for emergency back-up, so from then on we had in effect just one helicopter to be used by all of the geologists and paleontologists in camp. Although it did

Bill Breed adjusting the hitch on a Nansen sled; Leon Lambrecht of Belgium looking on. Weather permitting, we went back and forth by sled.

restrict those of us collecting fossil vertebrates, the restriction was not fatal to our program. We had plenty of work to do at Coalsack Bluff.

We did it day after day, weather permitting, going back and forth by sled. Getting to the exposures on Coalsack Bluff was not quite so simple as is implied by the preceding statement. Actually we drove our sled to the head of a col, where we parked it, and then walked in the rest of the distance—a half mile or so. The walking was not at all to my liking, because it was a choice either of crossing an ice slope that went down and down to perdition, it seemed, or struggling across loose scree above the slope. Bill and Jim would put crampons on their mukluks and cross the ice with the greatest of aplomb, but I could not quite face up to the long slippery slope with immense boulders and rocks at its foot. It terrified me. So I would struggle across the scree twice a day, slipping down two steps for every step I took forward and arriving at the fossil site thoroughly sweaty and thoroughly winded. It was truly frustrating when David Elliot would visit our fossil site; he would bound along or up the scree with the greatest of ease, stepping and leaping from boulder to boulder, while I floundered far behind.

All this time we were finding bones, but nothing truly definitive. I knew they were the bones of therapsid reptiles, but what kind of reptiles might they represent? One often needs a skull or a part of a skull to make an absolutely certain identification of a fossil reptile; generally speaking, the bones of the skeleton back of the skull are a little too generalized for close determinations. This is particularly true among some of the therapsid or mammal-like reptiles.

Every day I would hope for a skull, and my hopes were especially strong on the morning of December 4, out there on the slopes of Coalsack Bluff. "Why can't we find a skull of *Lystrosaurus,*" I said to myself. "These bones we have certainly look like *Lystrosaurus.*" I remember my thoughts on that day with unusual clarity, because it was the day when we did find a part of a *Lystrosaurus* skull. Jim Jensen found it during the course of the afternoon, but at the time he was not sure as to what he had.

When we got back to camp for the evening meal I started to go over the day's collection, cleaning and brushing specimens as they needed such treatment. Suddenly there was a good, identifiable *Lystrosaurus* in my hand—not a skull but a right maxilla with a tusk in place. There could be no doubt about it! I ran over to the mess hall where Jim was at the time and excitedly told him about his discovery. Jim immediately became as excited as I was, and our sentiments quickly spread through the Usarp hut. Everybody was chattering and exclaiming about the find, everybody except Isak Rust, who took it all very calmly .

"Why the fuss?" said Isak. "You have South African geology here, so why not South African fossils?"

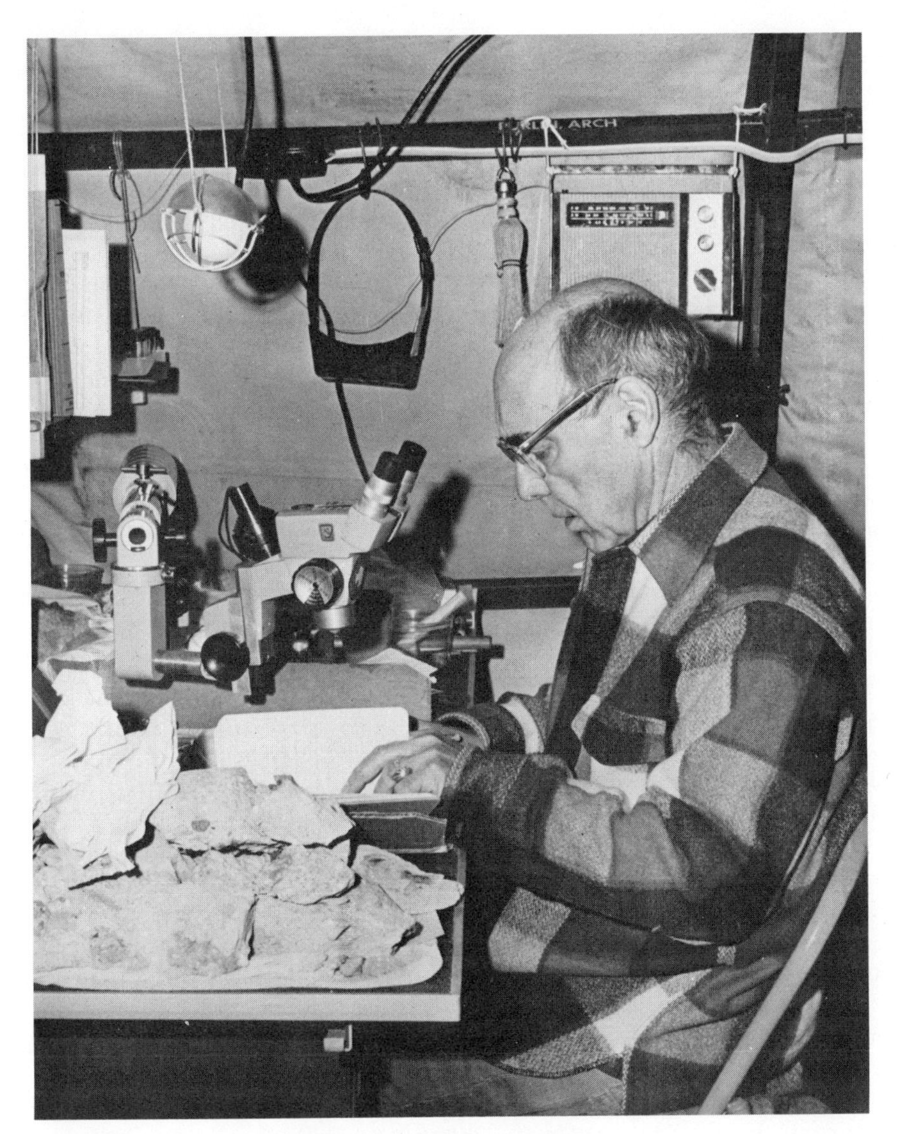

I started to go over the day's collection, cleaning and brushing specimens.

But Isak's *sang-froid* was not shared by the rest of us. We were keyed up by the discovery and could not quickly settle down.

It is sometimes wonderful how circumstances come together in real life in a manner that would not be attempted in fiction. That day I had wished for a *Lystrosaurus* and we had a *Lystrosaurus.* Then while we were at dinner word came in that a Hercules was on its way with Larry Gould, Grover Murray, Admiral Welch, in charge of the Navy contingent in Antarctica, and other distinguished visitors. The plane was due to land at about ten-thirty P.M.

Lystrosaurus already has been introduced in the chapter about Africa, and it appears again in the story of experiences in India. It need not be re-described at this place. Suffice it to say that this very typical Lower Triassic Gondwana reptile was now safely and unmistakably in our hands, and it seemed to us to be incontrovertible evidence pointing to the connection of Antarctica with Africa.

The plane landed as predicted, and very soon Larry Gould was just as excited as the rest of us. He immediately appreciated the significance of *Lystrosaurus* in Antarctica, and when he got back to McMurdo Base he flashed the news to the National Science Foundation in Washington, enthusiastically pronouncing our discovery "not only the most important fossil ever found in Antarctica but one of the truly great fossil finds of all

"Why the fuss?" said Isak. "You have South African geology here, so why not South African fossils?" Compare this picture with the photograph of the Karroo hill on page 167. Here is a bit of South Africa with *the snow and ice.*

time." This statement appeared on the front page of the *New York Times* of December 6, 1969, as part of a lead story written by Walter Sullivan describing the discovery of *Lystrosaurus* in Antarctica and outlining the former positions of Africa and the south polar continent as parts of Gondwanaland. It pleased me when later I saw the article, that Margaret's restoration of *Lystrosaurus*, which had been done for my book *The Age of Reptiles*, occupied a prominent place in the middle of the front page. Of course the story went out to news services all over the world, with the result that quite a few people who never before had heard of *Lystrosaurus* became acquainted with this strange Triassic reptile.

December 4, 1969, was a big day for us at Coalsack Bluff. From then on we collected fossils with more certainty as to what we were getting. I knew now that a great majority of the bones we had found and were finding were those of *Lystrosaurus*, which is also characteristic for fossils found in the *Lystrosaurus* zone of South Africa. We were, however, finding the remains of other reptiles and of amphibians as well. Much of what we were collecting at Coalsack Bluff would have to wait for detailed laboratory work and study in order to be sure as to the nature of the fossil assemblage from Antarctica. Nonetheless, it was looking more and more African in its several aspects.

There is not a great deal more to be said about our efforts in Antarctica. We continued for most of the time at Coalsack Bluff, but we did make some exploratory trips to other localities to see what we could see. What we saw was that the Fremouw Formation is widely exposed, and presumably richly fossiliferous, paralleling in those respects the *Lystrosaurus* zone of the Karroo rocks, so widely exposed in South Africa. We had planned to spend the second part of our field season at the junction of Shackleton and McGregor glaciers, 150 miles from Coalsack Bluff and on the other side of the Beardmore Glacier. But our late start in the field owing to bad weather, combined with helicopter problems, had set back our program to an appreciable extent. David and I consequently decided to forego Shackleton Glacier; to attempt work there would push us uncomfortably close to the end of the field season, with little time in which to accomplish significant results. The Shackleton-McGregor project was to be reserved for the following year, when James Kitching would lead the fossil collectors.

Early in January we packed our fossils in eight strong wooden boxes, ready for shipment to New Zealand and from there to the States. Immediately before our scheduled departure from the Coalsack Bluff camp, life was enlivened by the arrival of a Hercules carrying a committee from the House of Representatives and a contingent of Navy brass. Our Antarctic discovery had stimulated some congressional curiosity, so these members of the Committee on Science were coming to see if the taxpayers' money was

Early in January we packed our fossils in eight strong wooden boxes, ready for shipment to New Zealand and from there to the States. Looking pleased with the collection (from left to right), are Bill Breed, the author, and Jon Powell.

being well spent. They came to our mess hall, where David and I gave them talks about what we had been doing, and it appeared that they were well satisfied. The congressmen departed in a friendly mood and we departed with them.

That final take-off from our camp was a rough one—the plane skis banged across the sastrugi, or ice ridges, and it seemed to me that everything movable in the plane fell down before we were airborne. Finally we were off, and a few hours later had landed safely at Williams Field, McMurdo Base.

There were to be a few days at McMurdo before we left for New Zealand—days in which we made more secure our boxes of fossils and stenciled them for proper identification, days in which I watched two Coast Guard icebreakers far out in the sound fighting to open a navigable channel through the thick sea ice in order that freighters might make their annual quick dash into McMurdo with heavy supplies.

Finally there was to be our trip to the South Pole. The Admiral was so pleased by all the attention that *Lystrosaurus* had brought to Antarctica, which of course was favorable to the Navy, that he granted us a visit to the South Pole aboard one of the Hercules supply missions. We were up in good time and eagerly anticipating this very special experience. The flight was routine, and on the way we passed over our Coalsack Bluff camp although we did not see it, and after about three hours of flying we came in for the South Pole landing. It was so easy one could not help thinking of the contrast with Scott's harrowing journey of 1911–1912, a journey that ended in disaster and death.

The South Pole is in the middle of the great Polar Plateau, at an elevation of about 10,000 feet, most of those thousands of feet being ice. It is a flat, featureless ice field, at one time a great unbroken expanse of whiteness, now still a great expanse, but with the Scott-Amundsen Station making a tiny dot as one looks down from the air. We had a gourmet lunch at the station (the Navy makes a point of having one of the best cooks obtainable at the South Pole station), after which Bill affixed a paho feather at the actual South Pole, where an American flag flies, surrounded by a circle of flags representing all of the nations signatory to the Antarctic Treaty. It should be explained that a paho feather is a Hopi cultural item, a feather that is hung up in a house or at some special place as a blessing. Bill had brought the feather all the way from Arizona—a gift from one of our Hopi friends.

The day after our trip to the South Pole we left McMurdo for New Zealand. For me it was the end of a great adventure. Although our Antarctic experience had been most successful, and in many ways enjoyable, it was good to get back to a land of green vegetation everywhere, a land of trees and flowers and rushing streams and bright, emerald pastures. The thing I especially appreciated was darkness at night; it was refreshing to see sunsets, and to see the stars after months of constant daylight. On the day after our arrival in Christchurch, which was a Sunday, Bill and I went to the Botanical Garden, where I could enjoy the beauty of the flowers and the trees without any apprehensive thoughts of days to come—as had been the case when I first went to the Gardens before we flew to the Ice. It was a comfortable feeling.

Two days later we checked in at the airport for our flight back home, and there I hovered around the plane until I saw our boxes of fossils safely on board. A long flight back, with stops in Samoa and Hawaii, and finally a landing at the Alameda Air Naval Station near Oakland. Again I hovered about the plane until the boxes of fossils were unloaded. Margaret was there to meet me, in company with our old friends of American Museum and New York days, Rachel and John Nichols. We loaded all of the boxes in the car, causing it to sink down on its springs to an alarming degree, and giving us barely room to squeeze into the vehicle for the ride to their home.

A few days in Berkeley, where among other things I looked at *Lystrosaurus* fossils at the university, further to refresh my memory of this important reptile, and then we took a train to Flagstaff, with the boxes of fossils always under close supervision. On a day late in January, 1970, we arrived in Flagstaff with our fossils and were met by Ned Danson, the director of the Museum of Northern Arizona, and his wife, Jessica. We got the fossils to the museum, and my Antarctic adventure finally ended.

Yet my involvement with Antarctica was really only beginning. There were to be years ahead devoted in part to studying Antarctic fossils, to writing papers about them, and to attending Antarctic conferences.

16. ARIZONA

My involvement with Antarctica has continued largely because of the nature of paleontology. Once one is launched on a paleontological problem it can be a matter of years before the end is reached. In the first place, there usually is a considerable time lag between the collecting of fossils in the field, particularly the fossils of backboned animals, and the beginning of research upon these fossils. It is necessary to prepare them in the laboratory, the specialized and often tedious technological process of taking the fossils out of their plaster jackets if they are so enclosed, of chiseling them out of the encasing rock matrix if they are in a matrix, as they usually are, and of cleaning them and hardening them so that they can be handled. A difficult fossil may require months of such work. Only then are the fossils ready for study.

The study of fossils again is an extended process, involving a careful analysis of each specimen to determine its relationships, comparisons with other fossils or with recent animals, and the describing of the fossils under study. The description must be as accurate as is possible; detailed yet succinct, for once the description is published it is subject to searching review not only by one's contemporaries but also by generations of paleontologists in years to come. To make the description as useful as possible illustrations are essential—drawings and photographs. All of this takes added time.

The study of the Antarctic fossils has gone on, the work in the preparation and research laboratories exceeding many times over the time spent in collecting the specimens. Months have followed, one after another, and the months have extended into years.

My studies of Antarctic fossils have been multiplied by the acquisition of additional fossils collected under the leadership of James Kitching during a second field season. I was not a member of this second party, which worked in the Shackleton-McGregor Glacier area, where we had not been able to go during that first year. Needless to say, with James leading the fossil hunt success was assured. Some excellent fossils were found, supplementing and adding dimensions to the collections of the first year.

When James returned to California from his season on the Ice, Bill Breed and I drove out from Flagstaff in a carryall truck to meet him. We piled James and the fossils into the car, and brought them back to Flagstaff, where James and I had a preliminary look at some of the more accessible specimens. James then had to go home to Johannesburg. Nevertheless, he and I have collaborated on the descriptions of many of the Antarctic fossils—a long-range collaboration, it is true, but a successful one.

Now the Antarctic studies are coming to an end, my own, as well as the collaborations with James and with other paleontologists, notably John Cosgriff of Wayne State University, who helped me with the fossil amphibians. It should be mentioned that several years after the first two paleontological field trips in Antarctica John went down to the Ice to make still another collection of fossils—again in the Shackleton-McGregor area. It might be mentioned that on this trip and on the previous trip led by James,

Yet that earlier life will always be vivid in my memory. It was there that Margaret and I experienced the joys and the occasional sorrows of bringing up our family of five sons. Left to right, back row: David (2), George (1), Philip (3); front row, Daniel (4), Charles (5).

Brady Hall, the geology building at the Research Center of the Museum of Northern Arizona. This is my second scientific home, and a very pleasant home it is.

our third son, Philip, was there to lend his assistance as an experienced mountaineer. His help was much appreciated by the paleontologists and also by Antarctic geologists, so much so that Phil has now made five trips to the Antarctic.

I am now finding the time to think about and work on other paleontological projects: on the delicate little skeleton of a primitive ornithischian dinosaur from Upper Triassic beds about forty miles removed from Flagstaff, on other North American Triassic fossils, and on several book projects. This gives me more than enough to keep me busy, yet not so busy, I hope, as to preclude the enjoyment of our new, or rather not so new, life in Arizona—at the home we built a decade ago and at my quarters in the Museum of Northern Arizona. Such enjoyment includes fairly frequent visits from our immediate family, from children and grandchildren, and from friends as well.

From this perspective New York, New Jersey, the American Museum of Natural History, and Columbia University seem far away and long ago. Yet that earlier life will always be vivid in my memory. It was there that Margaret and I experienced the joys and the occasional sorrows of bringing up our family of five sons. It was there that I enjoyed the inestimable opportunities for paleontological growth afforded by four decades at the American Museum, my scientific home to be remembered with the greatest of affection. It was there that we lived a good life.

Now we lead a second good life, where I have a second scientific home, and a very pleasant home it is. Here I hope to stay during the years to come.

My office-laboratory in Brady Hall. There is much to look back on, and there are prospects ahead as well.

There is much to look back on, and there are prospects ahead, as well. I suppose I will be involved with fossils in one way or another until the day I die. Such is the pattern for paleontologists, because paleontology is more than a profession—it is a way of life. I cannot recall any paleontologists who have retired in the conventional sense of abandoning their previous activities, to spend their final days playing golf or shuffleboard, or sitting in the sun. They have devoted their lives to fossils, and fossils have a strong hold on their lives.

How does one quit? It is a hard question. I hope I will know when it is time to give up active research, when it is time to recognize that my efforts are more pitiful than helpful. I hope I can approach that time gradually, giving up fossils little by little but not completely, and spending more and more time at what seem inconsequential things. The recognition of this dividing line, or perhaps one should call it a dividing zone, a zone with very indefinite boundaries, is not at all easy. And this applies not only to paleontology, but to almost all aspects of human endeavor.

Colette, one of the most perceptive and delightful of twentieth-century writers, came to such recognition and expressed the problem as it concerned an author in her usual superb prose.

"But when does one stop writing? What is the warning sign? A trembling hand? I used to believe that the task of writing was like other tasks:

you put down the tool and shout with joy 'Finished'—and you clap your hands, from which rain down the grains of a sand considered precious. . . . It is then that in the pattern formed by the grains of sand you read the words 'To be continued. . . .' "

INDEX

(See also List of Illustrations, pp. ix-xi)